evasive
entrepreneurs
& the future of governance

"Entrepreneurs who skirt regulations that would block their ideas from coming to fruition aren't the greedy scofflaws they're sometimes portrayed as. Rather, they're engaged in a form of civil disobedience that can lead to better governance, and forward-looking public officials are taking that information to heart. In *Evasive Entrepreneurs and the Future of Governance*, Adam Thierer provides an important new way of reframing the debate around 'permissionless innovation,' with lessons for business, regulators, and everyone concerned with a fair and prosperous future."

—Virginia Postrel, author and columnist

"Thierer presents a bold and new concept that will disrupt how society manages disruptive technologies."

—Gary Marchant, professor of law at Arizona State University's Sandra Day O'Connor College of Law

"As our overly bureaucratized society threatens every day to stifle the innovation, discovery, and entrepreneurship that have for so long been regarded as a core element of American prosperity, Adam Thierer comes along with a book that shows us a better way. Blending real-life examples with the theories of social science, he shows why 'freedom to innovate is a moral imperative' both for individuals and society at large. If humanity is to have any future at all, it must be along the lines Thierer lays down in this book."

—Timothy Sandefur, vice president for litigation at the Goldwater Institute and author of *The Permission Society: How the Ruling Class Turns Our Freedoms into Privilege and What We Can Do about It*

"In *Permissionless Innovation*, Adam Thierer made a clear and powerful case against regulations that bind innovators. The book was oft-quoted, but also oft-ignored by regulators—'it is hard to teach someone something that it is not in their interest to learn.' But like the innovators whom he defends, Thierer courageously perseveres. In his hopeful new *Evasive Entrepreneurs and the Future of Governance*, Adam Thierer boldly argues that when onerous regulations persist, innovators can and should escape."

—Arthur M. Diamond, Jr., professor of economics at the University of Nebraska–Omaha and author of *Openness to Creative Destruction: Sustaining Innovative Dynamism*

"Adam Thierer bravely takes up arms against a sea of anti-technology histrionics, reminding us that the benefits of permissionless innovation far outweigh the costs. Though his calls for entrepreneurs to continue evading or even ignoring counterproductive and counterintuitive laws and regulations may sound radical, it's worth remembering that every great inventor in history followed that advice, to the vast benefit of society. This is a welcome counterattack to faddish technophobia. Thierer's common-sense recommendations will someday be heeded by policymakers, entrepreneurs, and consumers—hopefully before too much damage is done to the engine of innovation driving 21st-century life."

—Larry Downes, *New York Times* best-selling author/coauthor of *Unleashing the Killer App: Digital Strategies for Market Dominance, The Laws of Disruption: Harnessing the New Forces That Govern Life and Business in the Digital Age,* and *Pivot to the Future: Discovering Value and Creating Growth in a Disrupted World*

evasive entrepreneurs

& the future of governance

how
innovation
improves
economies
and
governments

—

Adam Thierer

Paperback ISBN: 978-1-948647-76-2
eISBN: 978-1-948647-77-9

Cover design: FaceOut Studio, Spencer Fuller.

Printed in the United States of America.

Library of Congress Cataloging-in-Publication Data available.

CATO INSTITUTE
1000 Massachusetts Avenue NW
Washington, DC 20001

www.cato.org

INTRODUCTION 1

1. WHY INNOVATION MATTERS 19

2. EVASIVE ENTREPRENEURIALISM: THEORY AND EXAMPLES 53

3. DISOBEDIENCE: THEN AND NOW 109

4. WHY EVASIVE ENTREPRENEURIALISM IS ON THE RISE 129

5. INNOVATION AS CHECKS AND BALANCES 159

6. HUMANISM, ETHICS, AND RESPONSIBLE INNOVATION 183

7. SOFT LAW AND THE FUTURE OF TECHNOLOGICAL GOVERNANCE 205

8. DEFENDING INNOVATION: A BLUEPRINT 241

9. CONCLUSION: THE CASE FOR RATIONAL OPTIMISM 265

POSTSCRIPT: EVASIVENESS DURING A PANDEMIC 277

ACKNOWLEDGMENTS 285

NOTES 287

INDEX 349

ABOUT THE AUTHOR 369

"The higher-ups have measures.
Those lower down have countermeasures."
—old Chinese saying

"Exit has an essential role to play in restoring quality
performance of government, just as in any organization."
—Albert O. Hirschman, *Exit, Voice, and Loyalty*

INTRODUCTION

This book argues that the freedom to innovate is important not only because it expands opportunities for economic growth and human flourishing, but also because entrepreneurial acts and the technological innovations they generate can help improve the quality of government policies and institutions.

Increasingly today, *evasive entrepreneurs*—innovators who don't always conform to social or legal norms—are using new technological capabilities to circumvent traditional regulatory systems, or at least to put pressure on public policymakers to reform or selectively enforce laws and regulations that are outmoded, inefficient, or illogical. Evasive entrepreneurs rely on a strategy of permissionless innovation in both the business world and the political arena.[1] They push back against "the Permission Society," or the convoluted labyrinth of permits and red tape that often encumber entrepreneurial activities.[2] In essence, evasive entrepreneurs live out the adage that "it is easier to ask forgiveness than it is to get permission" by creating exciting new products and services without necessarily receiving the blessing of public officials before doing so.[3]

Evasive entrepreneurs are taking advantage of the growth of various technologies of freedom, or what might also be labeled "technologies of resistance." These technologies are devices and platforms that let citizens circumvent (or perhaps just ignore) public policies that limit their liberty or their freedom to innovate or to enjoy the fruits of innovation.[4] We can think of this phenomenon as "technological civil disobedience." This term represents the technologically enabled refusal of individuals, groups, or businesses to obey certain laws or regulations because they find those laws or regulations offensive, confusing, time-consuming, expensive, or perhaps just annoying and irrelevant.

The technologies of freedom or resistance that facilitate evasive entrepreneurialism and technological civil disobedience include common tools such as smartphones, ubiquitous computing, and various new media platforms, as well as more specialized technologies such as cryptocurrencies and blockchain-based services, private drones, immersive technologies such as virtual reality, 3D printers, the Internet of Things, and sharing economy platforms and services. That list just scratches the surface,[5] however, and "the list of potentially disruptive technologies keeps getting longer."[6] "Inventions previously seen only in science fiction," a 2015 report from the World Economic Forum argues, "will enable us to connect and invent in ways we never have before."[7]

When innovators and consumers use tools and technological capabilities such as those to pursue a living, enjoy new experiences, or enhance the human experience, they often disrupt legal or social norms in the process. That disruption is not necessarily a bad thing. In fact, evasive entrepreneurialism can transform our society for the better because it can do the following:

- Help expand the range of life-enriching innovations available to society.
- Help citizens pursue lives of their own choosing—both as creators looking for the freedom to earn a living and as consumers looking to discover and enjoy important new goods and services.
- Help provide a meaningful, ongoing check on government policies and programs that all too often have outlived their usefulness or simply defy common sense.

For those reasons, I will argue that we should tolerate—and often even embrace—a certain amount of evasive entrepreneurialism and even a fair amount of technological civil disobedience. Defending successful acts of disruptive entrepreneurialism is easy after they occur; I seek here to defend the process that leads to those acts in the first place, which often receives less support. I will do so by making the case that the freedom to innovate is essential to human betterment for each of us individually and for civilization as a whole. That freedom deserves to be taken more seriously today. Finally, we should better appreciate how creative acts and the innovations they give rise to can help us improve government by keeping public policies fresh, sensible, and in line with common sense and the consent of the governed.

Risk Taking and Innovation Drive Progress and Human Flourishing

The normative case for evasive entrepreneurialism and the freedom to innovate begins with the fact that, throughout history, many innovations began as what were essentially illegal acts that were opposed by various government authorities, powerful institutions, or private interests.[8] Economic historian Deirdre N. McCloskey reminds us that "'betterment' and 'improvement' and especially 'innovation' were long seen in Europe as violations of God's will or as unsettling heresies."[9] That same negative thinking about innovation later infected other continents and cultures. With the passage of time, however, new technological tools and capabilities quickly go from being considered controversial to being commonplace—and even essential—because citizens usually embrace the improvements to their lives that those tools and capabilities enable.[10]

It often took bold acts by daring dreamers to get us to the point of acceptance, however. "Only those who will risk going too far can possibly find out how far one can go," the poet T. S. Eliot once noted. By their very nature, innovators and entrepreneurs break with tradition; they are "agents of change." They refuse to settle for the status quo. They imagine a different and better world, and they take risks to achieve their goals. "What entrepreneurs do," venture capitalist Vinod Khosla argues, "is they imagine what feels impossible to most people, and take it all the way from impossible, to improbable, to possible but unlikely, to plausible, to probable, to real!"[11] Sometimes that entails working at the margins of social norms and legal rules to change things. Other times it means striking at their very core.

We should be willing to tolerate a certain amount of such outside-the-box thinking because entrepreneurialism expands opportunities for human betterment by constantly replenishing the well of important, life-enhancing ideas and applications.[12] Entrepreneurialism and technological innovation are the fundamental drivers of economic growth and of the incredible advances in the everyday quality of life we have enjoyed over time.[13] They are the key to expanding economic opportunities, choice, and mobility.

This means that there is a *moral* dimension associated with innovation and entrepreneurialism, and with the economic growth and social improvement they help bring about. "Growth is valuable not only for our material improvement," Benjamin M. Friedman of Harvard University argues, "but for how it affects our social attitudes and our political institutions—in other words, our society's moral character, in the term favored by the Enlightenment thinkers from whom so many of our views on openness, tolerance, and democracy have sprung."[14]

This story isn't over—or at least it shouldn't be. Innovation can usher in new and better ways of doing things to help us improve the human condition even more.[15] If we hope to achieve still more moonshots—"radical but feasible solutions to important problems"[16]—then we must give innovation and entrepreneurs a wide berth, because in daring to dream of a better future, we open up a world of new opportunities for progress and prosperity.[17] We should be willing to do so even when those innovative acts sometimes challenge existing norms, institutions, and laws.

The Right to Earn a Living and to Innovate

The normative case for evasive entrepreneurialism also rests on the fact that it is often far too difficult for people to pursue an honest living today.[18] Citizens should have the right to freely pursue a living, not only to provide themselves and their families with income and sustenance but also to enjoy the freedom to engage in rewarding work.[19] Indeed, the freedom to pursue a living is simply an extension of the ideal of the pursuit of happiness that has long been a cherished American value.[20]

Our right to pursue happiness aligns with our corresponding rights to speak, to learn, and to move about the world we inhabit. In the United States, our constitutional heritage secured these rights and made it clear that we possess them simply by nature of being human beings. So long as we do not bring harm to others, we are generally free to act as we wish. Our rights to pursue happiness and to speak, to learn, and to move freely serve as the basis of more specific freedoms: the freedom to tinker and try or to innovate more generally.

Although many self-described humanist scholars vociferously critique each new technological development, in reality there are few things more human than acts of invention. At its root, innovation involves efforts to discover new and better ways of solving practical human needs and wants. The resulting tools and methods we create to better our lives are called technologies.

Unfortunately, many barriers exist to expanding innovation opportunities and our entrepreneurial efforts to help ourselves, our loved ones, and others. Those barriers include occupational licensing rules,[21] cronyism-based industrial protectionist schemes, inefficient tax schemes, and many other layers of regulatory red tape at the federal, state, and local levels.[22] We should not be surprised, therefore, when citizens take advantage of new technological capabilities to evade some of those barriers in pursuit of their right to earn a living, to tinker with or try doing new things, or just to learn about the world and serve it better.

Checking Government Power through Constant Innovation

Evasive entrepreneurialism and innovative activities can be valuable in another important way. In an age when many of the constitutional limitations on government power are being ignored or unenforced, innovation itself can act as a powerful check on the power of the state and can help serve as a protector of important human liberties.

Over the past century, both legislative and judicial checks and balances in the United States have been eroded to the point where they now exist mostly in name only. Although we should never abandon efforts to use democratic and constitutional means of limiting state power—especially in the courts, where meaningful reforms are still most feasible—the ongoing evolution of technology can provide another way of keeping governments in line by forcing public officials to constrain their worst tendencies and undo past mistakes. If they fail to do so, public officials risk losing the allegiance of their more technologically empowered citizenry.

Evasive entrepreneurialism is not so much about evading law altogether as it is about trying to get interesting things done, demonstrating a social or an economic need for new innovations in the process, and then creating positive leverage for better results when politics inevitably becomes part of the story. By acting as entrepreneurs in the political arena, innovators expand opportunities for themselves and for the public more generally, which would not have been likely if they had done things by the book. Ironically, by pushing up against social and legal norms in that fashion, innovators also often increase their chances of getting a fair shake from policymakers, who are forced to acknowledge a clear public interest in the fruits of expanded innovation opportunities.

But evasive entrepreneurialism and the freedom to innovate have even more profoundly salubrious effects on the republic once we conceptualize innovation as an important form of dissent. Dissent plays a vital role in society and especially in politics. Dissent challenges the status quo and encourages fresh thinking about what certain majorities regard as consensus, which may actually be in need of serious rethinking.[23] Disruptive activities rooted in forms of evasive entrepreneurialism and technologically enabled civil disobedience can make dissent even more visible and effective.

The very threat of occasionally opting out of broken or outmoded government policies can help shake up the stodgy status quo held by various individuals and bodies. We should not be afraid to speak up and challenge authority. Being entrepreneurial and innovative is another important way to make sure our voices are heard and our desires are respected.

Dissenting through innovation can help make public officials more responsive to the people by reining in the excesses of the administrative state, making government more transparent and accountable, and ensuring that our civil rights and economic liberties are respected. Political and judicial efforts aimed at checking government authority must continue, but innovation itself can also help ensure that government accountability and the consent of the governed retain some meaning in this country.

Living with the Pace of Technological Change

Although a powerful defense of evasive entrepreneurialism and technological civil disobedience can be built on such grounds, it is equally true that we are going to have to learn to live with a certain amount of this disruptive activity. The expansion of modern technological capabilities is rapid relative to the glacial pace of political change. This gap between the ever-expanding frontier of technological possibilities and the ability of governments to keep up with the pace of change is referred to as the "pacing problem," and it is a phenomenon explored throughout this book.[24]

The pacing problem is the great equalizer in debates over technological governance for two reasons.[25] First, with new technologies multiplying at such a rapid clip and building on top of one another in a symbiotic fashion, we live in an era of rapid-fire "combinatorial innovation."[26] In this environment, policymakers no longer have the luxury of procrastinating about many important governance decisions. In some cases, governments will catch up or at least slow the tide of technological change.

But new innovators and technologies will arise just as quickly as policymakers are coming to grips with other developments, creating what is referred to as a "competency trap."[27] In practice, this trap means that completely foreclosing innovative activities in most technological arenas is becoming increasingly challenging, costly, and unrealistic for governments.

The pacing problem manifests itself in another important way. As the public grows more familiar with, and reliant on, new technologies, individuals quickly assimilate those technologies into their lives and expect that more and better things will be around the corner. Once people have new devices and services and come to take them for granted, it becomes extremely hard to take them away. This relationship between technological change and societal expectations acts as an extraordinarily powerful check on the ability of regulators to roll back the clock on innovative activities. Once any particular technological innovation is out of the bottle, it will be increasingly difficult to stuff it back in.[28]

This book will examine the practical challenges that individuals, institutions, and governments will face as the pacing problem accelerates and technological civil disobedience becomes a more regular feature of modern life. Some suggestions will be offered for mitigating the downsides associated with these developments. But the primary focus here will be on providing a blueprint for progress that can help open the door to more opportunities for innovation and help significantly improve human welfare in both the short and long terms.

Constructive Government Responses

Governments are not completely powerless in the face of challenges associated with technological change. Evasive entrepreneurs and crafty consumers will have their share of rebellious *Star Wars* moments, but we can also expect many *Empire Strikes*

Back responses from policymakers. That political pushback will sometimes be swift and occasionally even effective in foreclosing evasive acts. We should never underestimate the power of government to use force to harass and intimidate individuals—especially entrepreneurs who think and act outside the box.

Policymakers face resource and knowledge constraints, however, and they need to consider sensible responses. Cracking down on creative minds and fast-moving technologies is a costly, time-consuming affair. Moreover, enforcement challenges will increase over time, primarily because of the unrelenting nature of the pacing problem as well as the growth of innovation arbitrage, or the movement of innovations to the jurisdictions where they are treated most hospitably. The combined effect of these trends will force public officials to think harder about the hassles of enforcing many of their existing laws and regulations.

It may seem counterintuitive, but the easiest way for governments to discourage technological civil disobedience will not be with ominous threats or formal sanctions aimed at eradicating such practices altogether. If public officials respond to legal evasion by doubling down on illogical policies and prohibitions, the public will not necessarily be any more likely to obey them. In fact, the public might instead increase dissent and disobedience in response. This result is called the "compliance paradox," which occurs when tighter rules simply lead to increased legal evasion and enforcement nightmares.[29] Thanks to the growth of technologies of resistance and increasing opportunities to engage in innovation arbitrage, the compliance paradox will become a more serious predicament for policymakers in coming decades.

Lawmakers and regulators need to consider a balanced response to evasive entrepreneurialism that is rooted in the realization that technology creators and users are less likely to seek to evade laws and regulations when public policies are more in line with common sense. Sensible innovation policy means that rules need to be even-handed and leave the door open to new competition and forms of technological change. Lawmakers need to work harder to clean up the morass of confusing policies they have already concocted and bring some sense to the regulation of existing and emerging technologies. But for reasons I will itemize throughout the book, I do not place much faith in the willingness or ability of legislative bodies to undertake those efforts. Vested interests—both inside and outside government—will fight to preserve the status quo at all costs.

Nonetheless, if lawmakers are willing to get the process started, smart public policies should focus on the most serious potential harms associated with new innovations and not get obsessed with far-fetched hypothetical scenarios. Good policy should be reasonable about violations of law at the margin and should treat technological disobedience as a learning opportunity; that is, a chance to recalibrate policies and bring them in line with new societal demands and technological realities. Of course, not all technological risks are equal, and some will require a more sophisticated governance strategy.

Although traditional hard-law approaches will always have their place, policymakers must think more entrepreneurially themselves and create more flexible, adaptive policy approaches for the new challenges they will face.[30] For anticipatory efforts, soft-law mechanisms—multistakeholder processes, industry best practices and standards, agency workshops and guidance documents, educational efforts, and more—can help address both ethical and technical governance questions without completely foreclosing innovation opportunities. Many government officials and agencies are already moving in this direction, recognizing that the combination of the pacing problem and evasive entrepreneurialism is eroding many hard-law policies and regulatory regimes.[31] Not everyone will be happy with this soft-law approach, but policymakers and other concerned parties need to realize that they have no choice but to undertake serious reform and adopt more flexible governance mechanisms if they hope to craft solutions that can keep pace with the technological changes we are witnessing today.

The common law will also continue to play an important role in addressing policy concerns in a reactive, remedial fashion. Common law mechanisms such as product liability, accident compensation, design defects law, failure to warn, breach of warranty, privacy torts, and trespass laws all have continuing importance. The common law evolves to meet new technological concerns and incentivizes innovators to make their products safer over time to avoid lawsuits and negative publicity.

Against Utopianism, Toward Pragmatic Change

As I hope to make clear, the approach documented and defended here is not rooted in any sort of grandiose, utopian theory of social or political change. I am not a crypto-anarchist who advocates revolutionary change via technologically enabled upheaval.[32] Nor will there will be any salvation-through-technology[33] pronouncements about innovation leading to the death of politics, the end of all regulatory shenanigans, or the complete demise of special interest influence on government.[34]

Serious political reform aimed at limiting the power of government over our lives and liberties is an extremely difficult, slow-going affair that requires multiple strategies—and a great deal of patience. The bureaucratic state has grown for decades and will likely continue to do so because it is difficult to stop the institutional forces aligned to preserve and extend its reach. But that does not mean that the so-called fourth branch of our government cannot be managed in other ways.[35]

The case I seek to make here may seem radical at first blush, but it is actually rooted in a fairly pragmatic vision and goal: *Beyond boosting economic growth and our standard of living, evasive entrepreneurialism can play an important role in constraining inefficient, unaccountable governmental activities that often fail to reflect common sense and the consent of the governed.*

Writing a half century ago, the economist and political theorist Albert Hirschman observed that "exit has an essential role to play in restoring quality performance of government, just as in any organization."[36] Innovative acts can be viewed as a type of exit, but ones that fall short of the more radical kind of exit the term conjures up in our minds. We need not call—as Thomas Jefferson once famously did—for repeated revolutionary acts to be undertaken every 20 years in an effort to "preserve the spirit of resistance" and keep government accountable to the people.[37] Along with other sensible governance methods and practical reforms, innovative acts can help us check governments' worst tendencies, reconsider the wisdom of the status quo in various contexts, and improve the quality of our political institutions and public policies—all without resorting to radical action.

This is why the freedom to innovate is so important and deserves a strong defense. This book seeks to provide that defense.

Map of the Book

The book opens with a simple, but often overlooked, question: Does innovation really matter? As Chapter 1 will show, innovation matters profoundly because it has been the primary driver of economic growth, human flourishing, and the long-term progress and prosperity of civilization. Innovation is fueled by continuous acts of entrepreneurialism and creative destruction, or what might more appropriately be labeled "innovative dynamism."[38]

Moreover, innovation allows individuals to live lives of their own choosing by making it easier to satisfy basic human needs. Thus, innovative activities are also worthy of a strong defense, even when they may sometimes be "evasive" in character.

That topic sets the stage for a discussion of what evasive entrepreneurialism means and how it has played out in various contexts. Chapter 2 offers several case studies and discusses how innovation arbitrage and jurisdictional competition will make technologically enabled disobedience even more likely in the future. That chapter also considers the prospects for evasion in sectors where technologies are "born free" of existing technocratic regulatory regimes versus "born captive," or likely to be burdened by existing regulations.

Chapter 3 steps back to consider how technological civil disobedience aligns with more traditional conceptions of civil disobedience. I will also seek to couch today's examples of disobedience in the broader American tradition of dissent and freedom of association that has deep roots in this nation's history. That chapter also explores the phenomenon of rule departure, or disobedient acts by government officials who sometimes choose not to enforce rules for various reasons.

Chapter 4 then explains why evasive entrepreneurialism and technologically enabled forms of dissent and disobedience are on the rise and likely to accelerate. Various factors and explanations will be explored, but the fundamental problem identified is that laws and regulations quite often defy common sense and typically fail to keep pace with new social and technological realities. This situation is caused by both the rapid pace of technological change (i.e., the pacing problem) and chronic failures within government itself. The result is less accountability and common sense in how government works today, which in turn encourages more evasion of broken or outmoded policies and processes.

That focus leads to a discussion in Chapter 5 of what may be my most controversial claim: With traditional legal and judicial checks and balances largely failing to keep policy up to date and protect important values and liberties, technological change itself may become the most important check on government power going forward. Critics on both the left and the right of the political spectrum may dispute this claim or be uneasy with it for various reasons. Nonetheless, I will argue that they should appreciate how these developments can have a positive effect on our government by helping make policymakers more accountable to the people and by bringing public policies more in line with common sense and modern realities.

Although evasive entrepreneurialism and technological change can have many positive benefits, they involve some undeniable tradeoffs. Innovation boosters cannot claim that the disruptive nature of creative destruction will be without challenges. Chapter 6 addresses some of the common objections raised by technology critics, who often rally under the banners of humanism and responsible innovation. Many of their concerns are valid and deserve a response—but not the innovation-limiting response that many of them desire. Slowing down or completely foreclosing entrepreneurial opportunities is almost never the wise approach. That idea leads in to the discussion in Chapter 7 about how more flexible and adaptive soft-law solutions are already being used to address these concerns. Soft law is not appropriate in every instance. Some technologies or technological processes give rise to more serious risks and deserve a more formal regulatory response. Chapter 8 grapples with those issues and begins sketching out a theory of technological harm that helps us decide when regulatory interventions are needed and which ones make the most sense.

Chapter 9 concludes by offering a variety of recommendations for how to both protect innovative acts going forward and improve government programs and procedures to encourage entrepreneurialism. Finally, other suggestions will be offered for how innovation advocates, universities, consumers, and entrepreneurs themselves can push for the freedom to both create and enjoy exciting, life-enriching innovations.

Key Terms Used in This Book

Compliance paradox: The situation in which heightened legal or regulatory efforts fail to reverse unwanted behavior and instead lead to increased legal evasion and additional enforcement problems.

Demosclerosis: Growing government dysfunction brought on by the inability of public institutions to adapt to change, especially technological change.

Evasive entrepreneurs: Innovators who do not always conform to social or legal norms.

Free innovation: Bottom-up, noncommercial forms of innovation that often take on an evasive character. Free innovation is sometimes called "grassroots" or "household" innovation or "social entrepreneurialism." Even though it is typically noncommercial in character, free innovation often involves regulatory entrepreneurialism and technological civil disobedience.

Innovation arbitrage: The movement of ideas, innovations, or operations to jurisdictions that provide legal and regulatory environments most hospitable to entrepreneurial activity. It can also be thought of as a form of jurisdictional shopping and can be facilitated by competitive federalism.

Innovation culture: The various social and political attitudes and pronouncements toward innovation, technology, and entrepreneurial activities that, taken together, influence the innovative capacity of a culture or nation.

Pacing problem: A term that generally refers to the inability of legal or regulatory regimes to keep up with the intensifying pace of technological change.

Permissionless innovation: The general notion that "it's easier to ask forgiveness than it is to get permission." As a policy vision, it refers to the idea that experimentation with new technologies and innovations should generally be permitted by default.

Precautionary principle: The practice of crafting public policies to control or limit innovations until their creators can prove that they will not cause any harm or disruptions.

Regulatory entrepreneurs: Evasive entrepreneurs who set out to intentionally challenge and change the law through their innovative activities. In essence, policy change is part of their business model.

Soft law: Informal, collaborative, and constantly evolving governance mechanisms that differ from hard law in that they lack the same degree of enforceability.

Technological civil disobedience: The technologically enabled refusal of individuals, groups, or businesses to obey certain laws or regulations because they find them offensive, confusing, time-consuming, expensive, or perhaps just annoying and irrelevant.

Technologies of freedom: Devices and platforms that let citizens openly defy (or perhaps just ignore) public policies that limit their liberty or freedom to innovate. Another term with the same meaning is "technologies of resistance."

1

WHY INNOVATION MATTERS

Before I discuss the growth of evasive entrepreneurialism and technological civil disobedience and provide more substantive examples, it is important to answer a more basic question: Why should we even care about innovation and the public policies governing entrepreneurial activities? This chapter considers some of the pushback against technological innovation and makes it clear how critics fail to account for the ways in which it has powered economic growth and human flourishing in a profoundly beneficial fashion throughout history.

For some people, asking why innovation matters will seem like a silly question, or one with a self-evident answer. Unfortunately, technological innovation is increasingly under attack from many politicians, academics, and social critics. In bestselling technology policy books that are assigned in university classrooms today, a common theme is evident: technology is something to be feared, and the benefits of innovation are dubi-ous or fictitious. "Cautionary voices on the risks of rapid techno-logical advancement are becoming louder," observes technology historian Calestous Juma in his 2016 book, *Innovation and Its Enemies: Why People Resist New Technologies.*[1]

That statement is hardly surprising, because a great many peo-ple prefer the status quo—whatever it may be for them or their organizations. When critics whip up panic about new technolo-gies, they are playing into people's worst fears and implicitly (sometimes quite *explicitly*) suggesting that the new and different are to be dreaded. But such status quo bias comes at an enormous cost—it means we will be giving up on the fruits of innovation because, as regulatory scholar Cristie Ford observes, "to a sub-stantial degree we regulate in response to dread."[2] If worst-case thinking and fear of the future come to dominate technology policy, we will not be able to discover new and better ways of ful-filling human needs and wants. That is why it is important to begin this book with a defense of innovation and entrepreneurial acts.

To be clear, technological innovation *does* involve tradeoffs and downsides, and regulation is sometimes needed to address serious risks associated with new technologies. Some innovators are nothing more than fraudsters, and some technologies by their very nature pose serious risk to life, limb, or property. Later chapters will discuss some of the more concerning risks associ-ated with innovation and how better governance systems must be devised to address them.

On balance, however, technological innovation has done far more good than harm to humanity's cause. Criticism of innova-tion needs to be grounded in reality and placed in context of what would have happened had we not innovated at all. Accordingly, this opening chapter discusses the swelling tide of anti-technology writing and advocacy and pushes back against the impulse to slow the wheels of progress. As we'll see, that impulse would create more harm to humanity than it would solve, because technological innovation has been the fundamental driver of improvements in human well-being over time.

The Rising Techno-Pessimist Tide

Scholars have noted that "public concern over technology is not entirely new."[3] It can even be traced to stories found in the book of Genesis that offer both positive (Noah's Ark) and negative (the Tower of Babel) perspectives on technology.[4] But technological criticism has, not surprisingly, ramped up significantly in recent times as technology has become far more intertwined with every aspect of modern life.

Opposition to technological innovation has many motivations. Innovation critics often sneer at "the cult of convenience"[5] and lament the supposed "paradox of choice,"[6] or the idea that too many choices overwhelm us. Most people regard convenience and choices as great benefits, but from the critics' perspective, the message is clear: more is less. Such skeptics prefer to slow things down and add friction to innovative processes in the name of protecting certain values or institutions that they fear technological change will erode.[7] They constantly lament the supposed anxiety or alienation of the masses, although it often seems like they are mostly airing their own grievances as opposed to seriously documenting those of the actual public.[8]

Regardless, tech critics like Evgeny Morozov advocate a "radical critique of technology" and a "radical project of social transformation"[9] that would evolve into a full-blown "degrowth movement."[10] As its very name implies, the degrowth movement questions whether growth is sustainable or even desirable.

Degrowth proponents also take issue with the idea that growth is synonymous with progress or human well-being.[11] Critics of this ilk insist that "there's nothing wrong with being a Luddite,"[12] because technology is dehumanizing and "will eliminate what it means to be human."[13] Critics also play into many people's unease surrounding the growth of new technologies like artificial intelligence (AI) and robotics and argue that these technologies will lead us to a "jobless future" run by machines and automation.[14]

A generation ago, engineering historian Samuel C. Florman noted that the technology critics of the Industrial Age were fond of portraying people as "helpless slaves" who were becoming "programmed machines" at the expense of new technologies.[15] Not much has changed since Florman made that assessment in 1976. If anything, today's anti-innovation screeds double down on that rhetoric and often take an even darker turn, predicting a veritable "existential threat" to the future of civilization.[16] Marc Goodman, author of *Future Crimes*, worries about our prospects for "surviving progress" and insists that "now is the time to completely reevaluate all that we take for granted in this modern technological world" before it is too late.[17] Another gloom-and-doom book from *The Atlantic* correspondent Franklin Foer warns that modern technology companies are taking us down a "terrifying trajectory" by eroding "the integrity of institutions" and "altering human evolution" to the point that a concerted resistance is needed to save humanity from technology.[18]

The tone of most of these tracts is relentlessly pessimistic, with dystopian dread dripping off each page as the authors wonder aloud about "how to keep technology from slipping beyond our control."[19] Technology is not a helpful servant to humanity in books such as these; it is instead a "dangerous master" to be feared and resisted.[20] Magazine stories, journal articles, and blogs about technology often incorporate the same dark themes.[21] References to Orwell and Huxley are common, and critics often insist that "surveillance capitalism"[22] will "lead toward a world turned into a Skinner box."[23] In other words, we are all just clueless lab rats being programmed to think and behave the way our corporate masters want, and the only benefit we get out of it, the critics say, is "cheap engineered bliss."[24]

Some of these modern tech critics may not even realize that they are articulating the same fears (and making some of the same gloomy Chicken Little predictions) that earlier generations of critics made many times before. In a best-selling 1970 book, futurist Alvin Toffler coined the term "future shock" to describe what he called "the disease of change" and the "real sickness from which increasingly large numbers already suffer."[25] Toffler argued this disease was brought on by the "shattering stress and disorientation that we induce in individuals by subjecting them to too much change too fast."[26] Had this disease really infected humanity to the extent that some tech critics imagined in the 1970s, one would think civilization would have already suffered catastrophic collapse. Humanity proved more resilient than those critics feared. Yet, roughly a half century later, books are still regularly published with similarly scary titles or phrases like "data smog,"[27] "information anxiety,"[28] "the tyranny of email,"[29] "dying for information,"[30] "the dumbest generation,"[31] "the end of reading,"[32] "digital barbarism,"[33] "digital vertigo,"[34] and "world without mind."[35]

Fear Entrepreneurs and False Equivalence in Tech Policy Debates

Many of the critics who use such apocalyptic rhetoric are acting as *fear entrepreneurs*. Fear entrepreneurs specialize in framing new technological developments as catastrophes to whip up panic and garner attention for themselves and their cause, whatever it may be.[36] One of the ways they do so is by using loaded language[37] and threat inflation, which has been defined as "the attempt by elites to create concern for a threat that goes beyond the scope and urgency that a disinterested analysis would justify."[38] At worst, critics engage in false equivalence, or the fallacy of inconsistency, by drawing comparisons between two situations or arguments that actually have very little in common. A few examples follow:

- Genetically modified organisms: Policy debates over genetically modified organisms (GMOs) are rife with threat inflation and loaded language such as "franken-food," the term GMO critics use to elevate fears about such products.[39] What is particularly astonishing about this example is that some of the most vociferous GMO opponents have very little scientific understanding of the GMO processes they decry in such extreme terms. In fact, a 2019 study in the journal *Nature Human Behaviour* reveals that "as extremity of opposition to and concern about genetically modified foods increases, objective knowledge about science and genetics decreases, but perceived understanding of genetically modified foods increases."[40] In other words, as a *Guardian* headline summarizes the findings, the "strongest opponents of GM foods know the least but think they know the most."[41] Many other studies document this sort of widespread ignorance about GMOs and other scientific and technical matters.[42] Despite this knowledge gap—or perhaps precisely because of it—tech critics will repeatedly fall back on threat inflation and coded language to scare the public and pursue innovation-limiting public policy objectives. Popular anti-GMO books use panicky titles such as *Seeds of Deception*[43] and *Altered Genes, Twisted Truth*.[44]

- Social networking safety: The rise of social networking sites in the mid-2000s provided another example of critics using threat inflation and coded language in an attempt to restrict a new technology, even though their claims lacked empirical evidence. Some politicians and attorneys general at the time claimed that sites like MySpace.com and Facebook represented a "predators' playground," implying that children could be groomed for abuse or abduction by visiting those sites.[45] Lawmakers even introduced provocatively titled legislation, the Deleting Online Predators Act,[46] which proposed a federal ban on access to social networking sites in schools and libraries to stop that menace. The measure would have indirectly censored many other online sites and services that had interactive functionality.[47] Serious academic inquiries into this danger, however, revealed that there was almost nothing to the predator panic. It was based almost entirely on threat inflation rather than credible evidence that such sites were being used to exploit children.[48] In fact, experts noted that social networking sites would be perhaps the worst places to attempt to groom children because predators' activities would be so easily traced and they would be brought to justice. Eventually, reason prevailed. Although the legislative effort to censor social networking sites via the Deleting Online Predators Act received an astonishing 410 votes in the U.S. House of Representatives in 2006, the measure failed in the Senate and did not see another vote. Nonetheless, similar technopanics continue to dominate each digital innovation that arises today.[49]

- Cybersecurity: Debates about digital security are often littered with false equivalence and threat inflation. It is not uncommon to hear pundits predicting the potential for a "digital Pearl Harbor,"[50] a "cyber cold war,"[51] a "cyber Katrina,"[52] or even a "cyber 9/11."[53] The security of interconnected digital and industrial systems is a legitimately serious matter deserving greater attention, and potentially even greater government action in some circumstances.[54] But such cyber vulnerabilities should not be equated with those historical incidents, which resulted in massive death and destruction. Meanwhile, popular books that touch on cybersecurity vulnerabilities include titles such as *Cyber War*[55] and *Click Here to Kill Everybody*,[56] which condition the ground of cybersecurity debates for even more extreme threat-inflation rhetoric from policymakers.

- Privacy: Debates over data collection and digital privacy provide still other examples of threat inflation and false equivalence in action. During a February 1, 2019, speech in Brussels, European Data Protection Supervisor Giovanni Buttarelli, who led data privacy efforts in the EU, encouraged stepped-up regulatory efforts on this front. During his remarks, Buttarelli referred to President Abraham Lincoln signing the Thirteenth Amendment to the U.S. Constitution on that date in 1865 and argued that "[w]e now face the challenge of abolishing digital servitude."[57] There can be no greater example of false equivalence than a casual comparison of digital data collection and the forced enslavement, servitude, and torture of humans. Buttarelli was only voicing the same sort of fanatical rhetoric seen in critical books about digital privacy including titles such as *Voluntary Enslavement*.[58] When threat inflation and false equivalence such as that statement run rampant in academic writing, eventually they come to infect the rhetoric of public policy debates and lead to technopanicky speeches and proposals.

One reason that pessimistic and apocalyptic themes pervade some nonfiction writing on innovation is that the writing is inspired by, and regularly borrows directly from, major fictional books, movies, and television shows.[59] Popular culture narratives about technology predict a future in which we humans are doomed to be enslaved by robots, controlled by malevolent AI systems, or, at best, have our brains turned into mush by machines, gadgets, or services that distract us into oblivion.[60]

In one sense, there is an obvious motivation for all these gloomy narratives: Bad news sells; good news doesn't.[61] "Pessimism has always been big box office," notes science writer Matt Ridley.[62] It is hard to sell a book or script with technology at the center of the plot unless the whole world goes to hell somewhere during the story. It would be wonderful if there were more celebration of innovation in popular culture;[63] alas, there instead seems to be a competition of sorts to see who can whip up more panic and grab more attention about how horrible technology is for us.[64] Meanwhile, journalists often use sensational, panic-inducing headlines and news reports when discussing breaking technological developments.[65]

It seems to be the case, as Ridley observes, that "[i]t's cool to be gloomy."[66] Another reason for this phenomenon, explains Clive Thompson, a contributor to *Wired*, is that "dystopian predictions are easy to generate" and "doomsaying is emotionally self-protective. . . . You seem like someone who has a richer, deeper appreciation for the past and who stands above the triviality of today's life."[67] In that regard, Ridley reminds us that philosopher John Stuart Mill observed this same tendency as far back as 1828 when he noted, "I have observed that not the man who hopes when others despair, but the man who despairs when others hope, is admired by a large class of persons as a sage."[68]

Everything New Is Feared until It Is Demanded

Today's gloomy tech critics and their fears about innovation are misplaced, and it is essential that such hysteria is debunked before real damage is done. Policy proposals inspired by the most recent episodes of the dystopian Netflix show *Black Mirror* or other science fiction doomsday dramas should be greeted with skepticism. Fictional depictions of the future are more often focused on challenging current cultural assumptions or offering a moral insight about the present than on accurately predicting technological progress and its consequences in the coming decades or centuries.[69] Dystopian hellscapes are an easy crutch to fall back on when looking to grab attention or seem enlightened,[70] but they do not reflect the actual story about technological innovation and how much it has done to improve the human condition.

Even the less extreme critiques of technological innovation are off base. For example, many tech critics wax nostalgic about some supposed good old days, or golden era, when various pastoral ideals regarding nature, farm life, or old traditions were honored and life was supposedly simpler and better.[71] "Demonizing innovation is often associated with campaigns to romanticize past products and practices," Juma correctly notes.[72] The problem with that "good old days" logic, he argues, is that "[o]pponents of innovation hark back to traditions as if traditions themselves were not inventions at some point in the past."[73] The things we hold dear in one era were likely ridiculed and viewed as scandalous, unconventional, or even criminal when they debuted. When looking back, we tend to normalize dramatic economic, cultural, and political shifts by thinking of them as incremental changes even though they were probably quite profound.[74]

Agricultural and food experimentation provides a good example of this. As historian Noel Kingsbury details in *Hybrid: The History and Science of Plant Breeding*, throughout history, whenever people started experimenting with new and better ways of breeding plants, critics started protesting about how it was some sort of sin against God or the natural order of things.[75] Once the benefits of each generation of new plant breeding techniques became evident, however, a new societal baseline was established and people began to demand those innovations—or they just took them for granted as if those innovations had always been around. But then the cycle begins anew with each food or plant innovation, as it has in recent years with GMOs.[76] Juma also recounts similar panics associated with a diverse array of food products from centuries ago (coffee), the past century (margarine), and then today (genetically modified salmon). The cycle never ends, and similar complaints can be expected to arise next about genetically modified designer foods,[77] lab-grown meat,[78] and 3D-printed food,[79] especially because many of those innovations have developed outside traditional regulatory regimes.

The same trends have played out in many other fields and sectors.[80] Following the rise of mechanization and industrialization, new panics have periodically popped up about how the latest machines of each era were going to automate all jobs out of existence.[81] The reality is that although some old skills and professions went away, many new ones replaced them.[82] Mechanization and automation have played an essential role in creating new and better jobs—and more of them over time.[83]

In my own research, I have documented the many fear cycles and technopanics that have come and gone for communications and media technologies.[84] These panics are timeless. Plato might be considered the first great tech critic, worrying as he did about the potentially deleterious effect that writing might have on the traditions of oral teaching and storytelling.[85] Humans quickly embraced writing and moved on. Much later, concerted opposition to the printing press developed because of fears about what easy replication of writings could mean for the Church, rulers, and the institutions of its day.[86] Society then came to revere books and the written word, only to turn its ridicule toward recorded music and broadcasting. More recently, critics have largely set aside fears about those old analog-era technologies and have turned their attention to digital-age villains: video games, computers, and now the internet, smartphones, social media platforms, and artificial intelligence.[87] In other words, "fear has gone hand in hand with technological advancements throughout history."[88] But somehow we always muddle through and come out better for it.[89]

We should not expect fear cycles or technopanics to subside, however. They will always be with us. As Chapter 6 will discuss, a great many self-declared humanists decry the dawn of each new technological era. They see the latest tools and technological trends as a sign of impending doom. Looking back over history, Richard Longworth of the Chicago Council on Global Affairs reminds us what we learn from opposition to the Gutenberg press and subsequent technologies:

> The monks of the day took an apocalyptic view of this disruptive technology, but there was more to come. James Watts' steam engine invented industry and cities. John Deere's plow killed the family farm. The Sears catalog (the Amazon of its day) undermined small-town stores. Trains

and steamships erased splendid isolation in favor a new connectivity. Don't get me started on Marconi, or radio, or TV.

All these innovations, like the digital revolution, had several things in common. They were engineering events that were economically successful. They drove out earlier technology, all the while changing how we live and think and relate to other humans. In other words, they forced us to re-think humanism, not only to adjust it to the new technology but to ensure that it eventually commanded this technology.[90]

We humans learned to harness our tools to make them serve humanity better. *Not perfectly, but better.* "Only a Pollyanna would claim that all these disruptions were for the best, or that they didn't bring immense human pain in their wake," Longworth observes. "But only an intellectual Luddite would throw up his hands and proclaim the death of humanism at the hands of technology."[91]

Unsurprisingly, these same fear cycles are unfolding today for drones, driverless cars, robots, virtual reality, genetic testing, and other technologies. Tech critics will vociferously protest each innovation and predict any number of calamities because of them. In fairly short order, however, most people—*including the critics themselves*—will likely assimilate these new technologies into their own lives and come to rely upon them. Today's technological boogeyman becomes tomorrow's technological necessity. Why? Because most technologies improve our lives in meaningful ways. We figure out how to make our tools serve us and society and how to mitigate the problems associated with them—often by building even more and better tools to do so.

What makes the nostalgic rhetoric about the past even more frustrating is that the proverbial good old days were not really so great.[92] In reality, historians have noted that the old days were "eras of misery," at least in terms of the basic standard of living for most people.[93] As we will see, by almost every important metric—life expectancy, overall health, wealth, education, and so on—we are vastly better off than our ancestors.[94] Return to the example of plants and agricultural science. As Kingsbury documents, most of us are alive today because of technological advances in plant breeding.

> Our access to food is so much more secure than that of our ancestors, our choices are incomparably wider, and our food is safer than ever before. Much of this security and choice are down to plant breeding. With the varied diet we have today it is difficult to imagine what it must have been like for our ancestors; their diet was often not just poor but incredibly monotonous. How many of us could imagine eating nothing but potatoes or rice for every meal for our entire lives, with vegetables dependent entirely on the season and meat or fish a rare luxury? The fact that many in the world still live like this, and suffer malnutrition as a result, is a powerful illustration of why we need plant breeding to continue to make advances.[95]

If plant breeding and agricultural innovation had been radically curtailed, these advances would not have occurred, and our collective well-being would have been worse off for it. We take many of these developments for granted today without even giving much thought to how they came about.

This is why technological innovation is so crucial—not just for food and sustenance, but for every field of science, and for every human need and want. Innovation is largely synonymous with human betterment and deserves a vigorous defense against attacks from critics. Once that case is made, it follows that keeping the door open to acts of entrepreneurialism is important

FIGURE 1.1

Share of world population living in poverty, 1820–2015

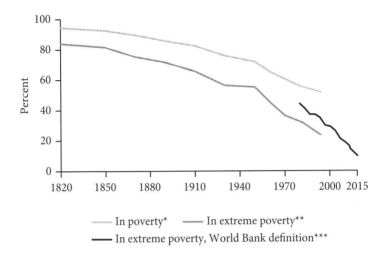

Sources: Max Roser and Esteban Ortiz-Ospina, "Global Extreme Poverty," *OurWorldInData. org*; and François Bourguignon and Christian Morrisson, "Inequality among World Citizens: 1820–1992," *American Economic Review* 92, no. 4 (Sept. 2002): 727–744, https://doi. org/10.1257/00028280260344443; World Bank (Povcal Net).

Note: *Less than $2 a day; **less than $1 a day; ***less than $1.90 a day. (All dollars are adjusted for inflation and purchasing power differences across countries.)

because those acts contribute to beneficial forms of social and economic change that can enrich our lives in deeply meaningful ways, such as decreasing poverty, as shown in Figure 1.1.

Innovation Drives Progress and Human Flourishing

Many tech critics would have us believe that innovation advocates are all just blindly worshiping our tools and developing a sort of "religion of technology."[96] The case for fostering a culture of entrepreneurialism and innovation is not about fetishizing any particular technological process or the latest shiny gadgets on the market. Nor does the case for innovation rest on the merits of any one particular innovation or entrepreneur. Everyone, and everything, has flaws and tradeoffs. Moreover, as later chapters discuss, some technologies can pose dangers to the public and deserve serious regulatory consideration.

Although some spotlight-seeking techno-utopians might be guilty of hero worship or excessive glorification of some tech gadgets or specific services, most innovation supporters have a very different goal in mind when defending technological innovation and entrepreneurial activities. The case for protecting creative minds and expanding entrepreneurial opportunities is premised on what has been shown time and again to improve the human condition in a profoundly beneficial fashion: ongoing trial-and-error experimentation with new and better ways of doing things.[97]

Economists, political scientists, and business theorists don't usually agree on much, but to the extent that they share a consensus about anything, it is that technological innovation is "widely considered the main source of economic progress"[98] and human prosperity more generally.[99] "The vast growth in wealth, health, and happiness is one of humanity's greatest successes," my Mercatus Center colleague Donald Boudreaux has noted.[100] Technological innovation has been the linchpin of that success story. "More than anything else technology creates our world," argues W. Brian Arthur in his book *The Nature of Technology*. "It creates our wealth, our economy, our very way of being."[101] The compounding nature of economic growth means that even an incremental reduction in the growth rate today will have profound consequences for the well-being of future generations.[102]

The evidence supporting this claim is voluminous. Countless economic studies and historical surveys have documented the positive relationships among technological progress, economic growth, and overall social welfare. For example, a 2010 Department of Commerce report from the Obama administration revealed that "technological innovation is linked to three-quarters of the nation's post-WWII growth rate," "innovation in capital goods is the primary driver of increases in real wages," and that "across countries, 75% of differences in income can be explained by innovation-driven productivity differentials."[103] In a recent white paper, James Broughel and I surveyed the substantial body of academic evidence proving this powerful relationship.[104]

Recent books by Hans Rosling,[105] Steven Pinker,[106] Robert Bryce,[107] and others[108] have also thoroughly documented these trends and shown that the historical evidence supports the unambiguous fact that "more people are living longer, healthier, freer, more peaceful, lives than at any time in human history," and that the "the simplest explanation is that innovation is allowing us to do more with less."[109] Here are some numbers that illustrate how, as a *Wall Street Journal* headline noted in early 2019, "The World Is Getting Quietly, Relentlessly Better":[110]

- Life expectancy: From 1800 to 2017, average life expectancy more than doubled, from just 31 years all the way to 72 years.[111]
- Extreme poverty: In 1800, 85 percent of humanity was surviving on less than $2 a day. By 1966, that number had fallen to 50 percent. As of 2017, just 9 percent of the world's population was living in such extreme poverty.[112] "Today almost everybody has escaped hell," Rosling asserts, noting that in just the past 20 years, "[e]xtreme poverty has dropped faster than ever in world history."[113] In fact, according to Max Roser, "on average, every day for the past 25 years 137,000 fewer

people were living in extreme poverty than the day before."[114]

- Infant mortality: Since 1960, child deaths have fallen from 20 million a year to 6 million in 2016.[115] During this time, the world population has grown from 3 billion to 7.7 billion. Infant mortality has plummeted, thanks to innovations in maternal health and neonatal medicine that have pushed the age of viability lower and allowed for treatment of previously deadly diagnoses.

- Vaccines: Not that long ago, immunizations against deadly diseases were uncommon. But from 1980 to 2016, the share of one-year-olds globally who got at least one vaccination jumped from just 22 percent to 88 percent.[116]

- Information and culture: With added leisure time, humans have come to enjoy a growing cornucopia of information and entertainment riches.[117] The number of new music recordings rose from almost zero in 1860 to 6.2 million per year by 2015.[118] New movies grew from nothing in 1900 to 11,000 per year by 2015.[119] And in just the past 40 years, internet access grew to reach almost 50 percent of the world's population by 2017, and 65 percent of people globally now have a mobile phone.[120] Much of this growth has been driven by the plummeting cost of computing. Data transit prices, for example, fell from about $1,200 per Mbps (megabits per second) in 1998 to just $0.02 per Mbps in 2017.[121]

These aggregate numbers tell an impressive story, but they fail to humanize technological innovation and help us understand why it has been so critical. One way to appreciate what technological innovation has meant for each of us is to look at how many hours of labor it took us to earn enough money to afford everyday goods and services in the past versus today.

Consider the transition from fire and candles as a source of illumination to electricity and light bulbs. Tim Harford, author of *50 Things That Made the Modern Economy*, provides a useful history lesson about how that move not only helped illuminate our world but also massively freed up time for us to do other things with our labor.[122] In the past, Harford notes, we needed to toil for 60 hours gathering and chopping wood over the course of a week to produce 1,000 lumen hours of light. "But that is the equivalent of one modern light bulb shining for just 54 minutes," and unfortunately, "what you would actually get is many more hours of dim, flickering light instead."[123] Today, we can use inexpensive light bulbs, and the result is not only more vibrant, long-lasting illumination but also huge labor savings. "The labour that had once produced the equivalent of 54 minutes of quality light now produced 52 years," he finds. The real price of light "has fallen by a factor of 500,000, far faster than official inflation statistics suggest."[124] Consider the many things that humans were able to accomplish with all that extra time. That's the real story of how innovation improves our lives and well-being.

The website HumanProgress.org tracks long-term improvements in human well-being and has documented similar cost of living improvements.[125] Table 1.1 uses data from the site to compare how much labor it took to purchase items in 1979 versus in 2015 and documents astonishing improvements.[126] The 50 to 100 percent reduction in the amount of labor needed to purchase these various goods represents a significant freeing up of leisure time to pursue other goals, whatever they may be for each of us. For example, in 1979, you had to spend a big chunk of your week (34 hours) earning enough to purchase a convection oven. In 2015, by contrast, you needed fewer than two hours of work to

TABLE 1.1

Hours of labor needed to purchase household items

	1979 (hours)	2015 (hours)	Percentage of reduction, 1979–2015
Refrigerator	75	36	52
Gas range	76	27	65
Dishwasher	33	13	61
Color TV	70	4.3	94
Treadmill	46	4.2	91
Vacuum cleaner	15	2.3	85
Convection oven	34	1.9	94

Source: Marian L. Tupy, "U.S. Cost of Living and Wage Stagnation, 1979–2015," HumanProgress.org, August 12, 2017.

buy it. In other words, in the course of just 36 years, innovation made it almost 100 percent easier to buy an oven for one's home and freed up our labor and money to accomplish other things.

Importantly, the improvements in access to these goods and in our standard of living are not just about cost or affordability. In each of these cases, *quality* improvements were as important, if not more important, than *price* improvements. Harford's example of poor-quality sources of illumination (fire and candles) relative to the light bulb is a dramatic illustration of how quality improvements are sometimes far more important to our living standards than data indicate.

Another example of quality improvement involves televisions. Although we can all appreciate just how much cheaper televisions have gotten during the past three decades, the real improvement has been the much higher-resolution screens that we now enjoy access to—many of which are much bigger (such as the large flat-screen TVs in our homes), while others are smaller and much more portable (including our phone and computer screens). This same trend is evident for many other technologies. Innovation and improvements happen along many dimensions.

Some comparisons of this type are now impossible thanks to new goods and services destroying old product categories or assimilating them into new ones. The paradigmatic example is the smartphone. Smartphones are still used to make phone calls, but our smartphones have now "morphed into the Swiss Army knife of gadgets"[127] and become "a multi-category killer."[128] We now use them as our cameras (replacing both photo and video cameras), portable music devices, televisions, gaming platforms, mapping and traffic navigation tools, web-surfing tools, notepads, email and messaging clients, payment tools, home automation agents, wearable fitness trackers, and even as watches, flashlights, compasses, and thermometers.[129] How does one measure the value of having to carry just one device instead of dozens? It is hard to know, but by offering us all those services in one easily portable device, our smartphones have clearly made many of our basic tasks more convenient.

What makes technological innovation such a powerful driver of human betterment is the combined power of trial-and-error experimentation and the social learning that results from it.[130] Defining "technology" can be tricky, but it generally comes down to "the application of organized knowledge to practical tasks"[131] and "the knowledge and instruments that humans use to accomplish the purposes of life."[132] By extension, technological innovation represents what Joseph Schumpeter described in 1935 as the "historic and irreversible change in the way of doing things"[133] or, more simply, "any new and better way of doing things," in the words of venture capitalist Peter Thiel.[134]

It is in that process of ongoing experimentation and learning new and better ways of doing things that we expand the overall universe of new goods and services, as well as opportunities and incomes. Over the long term, the virtuous cycle produced by this ongoing learning process becomes the central ingredient that helps countries raise their overall standards of living for their citizens, improve public health and knowledge, and achieve greater progress and prosperity.

There Can Be No Innovation without Entrepreneurialism and Risk

Entrepreneurialism and the creative destruction it brings about are crucial to this process, because it is only through risk taking and trial and error that we can continuously replenish the well of important ideas and innovations.[135] "The history of the human race would be dreary indeed if none of our forebears had ever been willing to accept risk in return for potential achievement," notes H. W. Lewis, an expert on the role that risk taking plays in fueling success and progress.[136]

Entrepreneurs are crucial to that learning process because they "repeatedly inject novelty into the economy"[137] and help ensure that stagnation does not take hold. The economist William Baumol once described entrepreneurs as individuals "willing to embark on adventure in pursuit of economic goals."[138] By being adventurers, entrepreneurs drive innovation and their innovative efforts and then drive growth and propel society forward.[139] "If we seek to explain the success of those economies that have managed to grow significantly compared with those that have remained relatively stagnant," Baumol argued, "we find it difficult to do so without taking into consideration differences in the availability of entrepreneurial talent and in the motivational mechanism which drives them on."[140] So central is the entrepreneur to the story of human progress that in his *Brief History of Entrepreneurship*, Joe Carlen argues:

> entrepreneurship stands alongside the other perennial elements of the human condition as one of history's prime movers. It has not only helped shape the kingdoms, empires, and civilizations of our world but, in many instances, it was entrepreneurship that provided the initial impetus behind their creation.[141]

This is why I believe a powerful moral case can be made for promoting entrepreneurialism and the freedom to innovate because of their connection to long-term economic growth and human prosperity.[142] As Edd S. Noell, Stephen L. S. Smith, and Bruce G. Webb argue in a recent book on the importance of economic growth:

> Economic growth was the key that transformed societies from dire poverty to prosperity and well-being. It has brought billions of people out of poverty and holds the promise of sustaining even higher levels of human flourishing if it continues. . . . [B]ecause growth is foundational to material well-being, it is also fundamentally a moral issue. People who care about human well-being, and who care about the poor, should promote growth. Devising policies to promote and sustain growth, in rich and poor countries, is a moral imperative.[143]

If we can generally agree that economic growth is a moral imperative, and that technological innovation and entrepreneurialism are the primary drivers of economic growth, then we can appreciate why the general freedom to innovate is also a moral imperative.

Economic growth and entrepreneurial activities also bolster a pluralistic society and have other powerful positive moral consequences. "Economic growth—meaning a rising standard of living for the clear majority of citizens—more often than not fosters greater opportunity, tolerance of diversity, social mobility, commitment to fairness, and dedication to democracy," notes Benjamin Friedman.[144]

Finally, and perhaps most important, innovation and entrepreneurialism are inextricably connected to human freedom and personal autonomy more generally. Innovation and economic growth are important because they allow us to live lives of our own choosing[145] and enjoy the fruits of a pluralistic society.[146] As Rosling rightly argues, the goal of expanding innovation opportunities and raising incomes "is not just bigger piles of money" or more leisure time.[147] He writes, "The ultimate goal is to have the freedom to do what we want."[148]

Another way to look at these beneficial effects is to consider the role technology and innovation play at every stage of Maslow's pyramid, which describes a five-level hierarchy of human needs. The most basic needs are physiological (survival needs like food and water) and then safety related (shelter, stability, etc.). Until the time of the Industrial Revolution, very few people had the luxury of looking beyond those basic survival and security needs. Feeding themselves and their families took up most of their time. Protecting themselves and their families—from animals, the elements, or human adversaries—was equally challenging in many circumstances. Technological innovation changed that by allowing us to satisfy those more basic survival and safety needs. As those needs were met, the possibility opened of satisfying higher needs on Maslow's pyramid, such as belonging, self-esteem, and self-actualization.[149]

For example, how much time and money did the average person in medieval Europe have to devote to the arts and entertainment? How did people enjoy their leisure time? Where did they go to relax on vacation? These questions are ludicrous because none of these things was feasible for anyone except a handful of elites. Today, technological innovation has helped an ever-expanding number of people escape dire poverty and a constant fear for their survival. As it has done so, technological innovation has allowed us to explore and satisfy other needs and desires, including many that our ancestors could have only dreamed about.[150] Self-actualization is not something one has a lot of time to worry about when one is toiling in a field trying to figure out how to put the next meal on the table.

Critics will retort that technology has also opened up new problems for us at the same time. There is some truth to that, and we should think hard about how to address those problems. Chapters 6 and 7 will do so. By any fair and sensible accounting, however, the positives associated with technological change have greatly outweighed the negatives. Moreover, as innovation has opened up a broader world of possibilities, the very notion of what we consider a problem has been watered down. When some complain about problems such as information overload or having too many choices, it is vital to put such problems in historical perspective. If the primary indictment of technological innovation is that it has inundated us with too much information or too many options, those are good problems compared with the more serious problems our ancestors faced.

The crucial takeaway here is that technological innovation has not only provided for our most basic needs as a species and expanded the material wealth and well-being of civilization, but also simultaneously expanded our ability to live lives of our own choosing. Robert Friedel, author of *A Culture of Improvement: Technology and the Western Millennium*, has put it best:

> Technology and the pursuit of improvement are ultimate expressions of freedom, of the capacity of humans to reject the limitations of their past and their experience, to transcend the boundaries of their biological capacities and their social traditions.[151]

Any exploration of the ethical consequences of technological change cannot be divorced from these realities. As Chapter 6 will note, technology has also expanded our moral universe by opening our eyes to the plight of others across the globe and giving us ways to address their needs. In this sense, innovation and humanism are again fundamentally intertwined.

Iteration, Innovation, and Moonshots

Innovation scholars often debate the differences between types or degrees of innovation. Sometimes they contrast iteration (doing old things better) with innovation (doing entirely new things) and debate which is more important.[152]

Both are important, of course. It is worth remembering that "modern versions of certain particular technologies do descend from earlier forms"[153] and that iterative processes and technologies improve human welfare in many meaningful ways.[154] Past technologies provide the building blocks for present ones.[155]

But when we think about what really moves the ball forward for society, it is truly transformative innovations, or what some call moonshots, that present the most exciting prospects for humanity. After all, iteration can only occur once entirely new products and services have been created. The well of important ideas and inventions must be constantly replenished if we hope to improve the health and happiness of our species in a meaningful way.

Most people intuitively understand and appreciate what technological innovation has done to improve their lives.[156] In late 2017, the Pew Research Center conducted a poll asking, "What would you say was the biggest improvement to life in America over the past 50 years or so?" As seen in Figure 1.2, most respondents (42 percent) said technology had contributed more than any other factor.[157] That was three times as many people who responded "medicine and health" (14 percent), which was the

FIGURE 1.2

Looking back: biggest improvement to life in the past 50 years

Percentage of U.S. adults who said the biggest improvement to life in America over the past 50 years or so was related to . . .

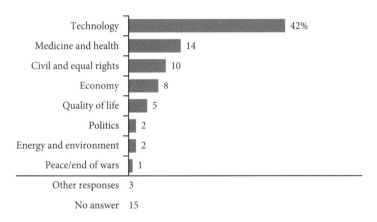

Source: Mark Strauss, "Four-in-Ten Americans Credit Technology with Improving Life Most in the Past 50 years," Pew Research Center (Fact Tank), October 12, 2017.

Note: Verbatim responses are coded into categories; figures are based on combining related codes into net categories. Figures add to more than 100 percent because multiple responses were allowed.

second-place answer. Incidentally, "politics" came in sixth place, with just 2 percent of the vote.

When we talk about disruptive innovation, moonshots, deep technologies, and technological miracles, what do we mean?[158] These terms can be overused, clichéd, and taken to unreasonable extremes.[159] "Moonshot" is a particularly confusing term because, for many, it conjures up massive government-funded initiatives, such as the Apollo program that initially gave rise to the term.

Nonetheless, these terms retain real meaning and significance, and they need not require any government initiative. Boudreaux defines moonshots more generically as "radical but feasible solutions to important problems."[160] Similarly, "deep technologies" are innovations that are "built on tangible scientific discoveries or engineering innovations," argues Swati Chaturvedi of investment firm Propel[x].[161] "They are trying to solve big issues that really affect the world around them," he writes.[162] Other terms that describe the same thing are "breakthrough technologies," "radical innovations,"[163] or "seismic innovation."[164]

Regardless of what we call them, the question we need to ask is: what big, life-changing innovation opportunities are we missing out on that could help us achieve significant improvements in human welfare and happiness?[165] How can our society, our thought leaders, and our public policies encourage more outside-the-box thinking that promotes bold ideas and innovations that hold the potential to massively improve the human condition?

Many of the innovations discussed throughout this book hold such promise—drones, driverless cars, genetic editing, 3D printing, AI and robotics, cryptocurrency and blockchain technologies, and more. Today's moonshot technologies hold exciting potential, but the yet-to-be-invented technologies of the future are what should excite us even more. We cannot rest on past successes; we must look to the future.

Free Innovation and Social Entrepreneurs

Importantly, innovation today springs from many different sources, including a multitude of noncommercial efforts of a grassroots or household origin.[166] MIT economist Eric von Hippel refers to this creation as a *free innovation*, which he defines as "a functionally novel product, service, or process that (1) was developed by consumers at private cost during their unpaid discretionary time (that is, no one paid them to do it) and (2) is not protected by its developers, and so is potentially acquirable by anyone without payment—for free."[167] In earlier writings, von Hippel coined the term *user entrepreneurialism* to describe closely related activities.[168] Others have defined user entrepreneurship as "entrepreneurship by individuals who create an innovative product or service because they need it for their own use and subsequently found a firm to commercialize their innovation."[169]

Social entrepreneurialism is a closely related concept.[170] It has been defined as "innovative, social value creating activity that can occur within or across the nonprofit, business, or government sectors."[171] As the name implies, some sort of social goal or vision drives social entrepreneurship. Social entrepreneurial activities are not typically in pursuit of compensation or profit, but that need not always be the case, and "the distinction between social and commercial entrepreneurship is not dichotomous, but rather more accurately conceptualized as a continuum ranging from purely social to purely economic."[172] For example, an individual or a group that establishes an online giving campaign to help others in need in their community would be a social entrepreneur. Later, however, they might attempt to convert that charitable effort into a formal business. That change would represent a move away from pure social entrepreneurialism and further down the continuum toward traditional profit-seeking entrepreneurialism.

Free innovations, user entrepreneurialism, and social entrepreneurs are a major part of the story told in this book because they are expanding so rapidly and creating novel problems for technological governance efforts. The enforcement of most technology laws and regulations hinges upon commercial transactions. Amazingly, functionally equivalent entrepreneurial activities are sometimes regulated in very different ways depending on whether just a few dollars are charged for a service.[173]

We will return to this strange anomaly throughout the book and discuss its implications for various bodies of law. For now, it is important to acknowledge that (a) free innovations and social entrepreneurs promise to change the world in major ways going forward as individuals and groups take advantage of the same ubiquitous, low-cost technologies that for-profit actors do, and (b) a fluid continuum exists of free versus for-profit innovation and social versus economic entrepreneurialism. In other words, innovation comes in many flavors, and that fact will complicate technological governance efforts.

Getting Innovation Policy Right Is Important

Once we understand that "a symbiotic relationship exists between technological innovation and human flourishing"[174] and appreciate the great potential for specific applications like those described earlier to result in meaningful improvements in human welfare, it becomes clear that *opening up new innovation opportunities should be a major public policy priority*. In practice, policymakers should give innovators greater freedom to experiment with new and different ways of doing things, even when those efforts prove to be highly disruptive. Moonshots and other entrepreneurial endeavors aren't as likely without public policy support for a culture of experimentation and risk taking.

In my previous book, *Permissionless Innovation: The Continuing Case for Comprehensive Technological Freedom*, I described the philosophical conflict of visions between two very different governance ideologies—permissionless innovation and the precautionary principle—and discussed how the tension between them dominates academic and policy debates about emerging technologies.[175] Permissionless innovation refers to the idea that experimentation with new technologies and business models should generally be permitted by default. It comes down to a general acceptance of change and risk taking. As the old saying goes, "nothing ventured, nothing gained." To form a governing vision from this idea, we can think of it as the "innovation principle," which holds that "the vast majority of new innovations are beneficial and pose little risk, so government should encourage them," in the words of Daniel Castro and Michael McLaughlin of the Information Technology and Innovation Foundation.[176] Permissionless innovation or the innovation principle means that technological change will be greeted with humility and an open mind and that patience and regulatory forbearance will be the guiding values when calls to stop progress are heard.

Those calls to slow or stop progress often come from pundits and policymakers who advocate the "precautionary principle" as the first-order response to technological change. The precautionary principle represents the antithesis of the innovation principle. It holds that innovations are to be curtailed, or potentially even disallowed, until the creators of those new technologies can prove that they will not cause any theoretical harms.[177] Precautionary policymaking is rooted in a mentality that says it is better to be safe than sorry. This mentality discourages risk taking and trial-and-error experimentation in favor of the upstream governance of new tech. Precautionary principle advocates will often insist that—if new technologies must be introduced into society at all—they must first undergo extensive evaluation by government officials to determine if they are in the public interest.[178] Stability and preservation of past laws, values, or business methods are the dominant values in the precautionary worldview, and these views are especially prevalent among academic critics and risk-averse politicians.[179]

The first goal of my previous book was to offer a *descriptive* account of how these two governance visions were at odds in policy battles over a wide range of technologies, including the sharing economy, autonomous vehicles, commercial drones, artificial intelligence, the Internet of Things, big data, Bitcoin, biotech, 3D printing, and advanced medical and health technologies.

As the title of my earlier book made clear, however, I also sought to make the *normative* case for permissionless innovation as the superior default regime for technological governance. My argument for rejecting the precautionary principle as the default came down to belief that "living in constant fear of worst-case scenarios—and premising public policy on them—means that best-case scenarios will never come about. When public policy is shaped by precautionary principle reasoning, it poses a serious threat to technological progress, economic entrepreneurialism, social adaptation, and long-run prosperity."[180]

Heavy-handed preemptive restraints on innovative acts have such deleterious effects because they raise barriers to the entry of new ideas and inventions into society, increase compliance costs, and create more risk and uncertainty for entrepreneurs and investors. Progress is impossible without constant trial-and-error experimentation and entrepreneurial risk taking.[181] Thus, the unseen costs of forgone innovation opportunities are what make the precautionary principle so troubling as a policy default. Without risk, there can be no reward. Scientist Martin Rees refers to this truism about the precautionary principle as "the hidden cost of saying no."[182]

More generally, risk analysts have noted that the precautionary principle "lacks a firm logical foundation"[183] and is "literally incoherent"[184] because it fails to specify a clear standard by which to judge which risks are most serious and worthy of preemptive control. Moreover, regulatory policy experts have criticized the fact that the precautionary principle "may be misused for protectionist ends; it tends to undermine international regulatory cooperation; and it may have highly undesirable distributive consequences."[185] Specifically, large incumbent firms are almost always better able to deal with rigid, expensive regulatory regimes or, worse yet, they can game those systems by "capturing" policymakers and using regulatory regimes to exclude new rivals.[186] (Regulatory capture is a problem we will revisit throughout the text.)

To the maximum extent possible, then, policymakers should make permissionless innovation their lodestar when thinking about the future of innovation. Our default principle for innovation also has implications in the race for global competitive advantage.[187] "Invention and entrepreneurship are at the heart of national advantage,"[188] making the choice for countries clear: innovate or perish.[189] Those countries and continents that make more innovation opportunities available to their citizenry will prosper; those that foreclose or limit those opportunities—or reward political entrepreneurs who seek special privileges—will stagnate.[190] "If a nation wishes to promote higher levels of domestic entrepreneurship in both the short and long run, top priority should be given to reducing barriers to entry for new firms and to improving overall institutional quality (especially political stability, regulatory quality, and voice and accountability)," Dustin Chambers and Jonathan Munemo note in a 2017 Mercatus Center report on the effect of regulation on entrepreneurial activity.[191]

In Chapter 8, I will outline a variety of reforms that could help make this goal a reality. I will also note that, unfortunately, the prospects for serious reform are quite limited. Sensible legislative and regulatory policies are notoriously difficult to achieve when so many people and organizations have an interest in preserving the status quo.

Entrepreneurs and average technology users have come to recognize this problem and, as a result, many of them are increasingly looking for ways around the system. The next chapter explores how and why they are behaving in such an evasive fashion to advance change, even when politics sometimes prevents evasive entrepreneurialism or makes it more difficult.

2

EVASIVE ENTREPRENEURIALISM:
THEORY AND EXAMPLES

There exists today a diverse array of new technologies of freedom, and they are growing in volume and capability. Innovators and average citizens are using the new technological tools and capabilities to express their creativity and develop exciting new gadgets and services to improve their lives and the lives of others. Using the new tools and platforms, evasive entrepreneurs sometimes run up against laws and regulations that make their efforts and ability to earn a living much more difficult. In the process, many of those creators end up engaging in acts of technological civil disobedience.

This chapter begins with a general discussion of evasive entrepreneurialism and identifies some of its core elements. It will also explore the relationship between evasive entrepreneurialism and permissionless innovation before addressing the related phenomenon of innovation arbitrage, which was described in the Introduction. This chapter will then offer several examples of evasive entrepreneurialism and technological civil disobedience in action before considering the prospects for evasive strategies in various sectors.

Regulatory Entrepreneurs

Economists and political scientists have documented the growth of evasive entrepreneurs as well as regulatory entrepreneurs. Scholars have defined evasive entrepreneurialism as "profit-driven business activity in the market aimed at circumventing the existing institutional framework by using innovations to exploit contradictions in that framework."[1] In economic terms, evasive entrepreneurs are attempting to maneuver around existing firms who have raised their rivals' costs through public policies that protect incumbents and their business models.

Regulatory entrepreneurs have been defined as innovators who "are in the business of trying to change or shape the law" and are "strategically operating in a zone of questionable legality or breaking the law until they can (hopefully) change it."[2] To the extent that there is a distinction between *evasive* and *regulatory* entrepreneurs, it might be in the degree of knowledge and intent of the entrepreneur regarding existing public policies. Regulatory entrepreneurs understand better how public policies affect their innovative activities, and they actively set out to change those policies directly through their actions. Evasive entrepreneurs could perhaps represent a slightly broader universe of innovators, not all of whom necessarily understand exactly how public policies affect their activities or, even if they do, are not necessarily seeking to change those laws or regulations (although most of them probably would like to do so).

To keep matters simple, I have already proposed a simpler definition of evasive entrepreneurialism as innovative activities that do not always conform to social or legal norms. Evasive entrepreneurs engage in a sort of political entrepreneurship—but of a different sort from the traditional meaning of the term. Most literature on political entrepreneurs identifies them as individuals and organizations that seek specific rewards from political institutions and interactions.[3] In seeking to secure privileges from governments[4] through "alertness to previously unnoticed rent-seeking opportunities,"[5] they are engaged in what other scholars often refer to as unproductive entrepreneurship.[6] Economists use the term "rent seeking" to describe the way individuals or groups seek to obtain special protections or privileges through a political process. These privileges can include tax breaks, subsidies, entry barriers, or other affirmative benefits. Chapter 4 discusses the problems associated with rent seeking in more detail.

By contrast, evasive entrepreneurs or regulatory entrepreneurs are not out to gain privileges through their activities. Instead, they are generally more interested in changing or sometimes just avoiding laws and regulations in an attempt to gain more operational freedom. George Mason University economists Peter Boettke and Christopher Coyne note,"[i]n general, as rules become more burdensome and raise the costs of interaction, one should expect economic actors to invest more resources in avoiding those rules."[7]

The general definition of evasive entrepreneurialism that I set forth earlier—innovative activities that don't always conform to social or legal norms—makes it clear that evasive entrepreneurialism need not always be in pursuit of profits, although that is often the case. Sometimes average citizens work either independently or together to create important innovations, or use technologies to pursue a cause, without necessarily seeking to make money in the process.

One entertaining example involves something as mundane as road repairs. With cities across America often failing to do basic infrastructure repairs, pothole vigilantes are taking matters into their own hands by filling in potholes.[8] These vigilantes use walkie-talkies to coordinate their work and later upload photos and videos of their work on Instagram and YouTube. Some of the pictures and videos highlight how they use plants and artistic decoration to make the potholes look nice. Those vigilantes are engaged in a form of evasive entrepreneurialism, even if none of them is seeking a profit. As discussed previously, such innovators are often called social entrepreneurs and their work is known as free innovation.

Whether they are seeking monetary profit or not, I will continue to group such innovators under the banner of evasive entrepreneurs, even though I am well aware that profit-seeking activity has traditionally been considered one of the core components of entrepreneurialism. In the next chapter, I will spend more time discussing whether for-profit activities should be grouped under the banner of technological civil disobedience.

In a sense, evasive entrepreneurialism has always been with us. One could even argue that, by their nature, entrepreneurs are inherently "kindled by a streak of rebellion"[9] and in "the business of breaking the settled mold."[10] Accordingly, they regularly act in an evasive fashion by pushing back against social norms, marketplace standards, entrenched business models, and even government rules. Economists Joseph Schumpeter[11] and Israel Kirzner[12] did pioneering work on entrepreneurial theory, highlighting how entrepreneurs act as paradigm busters or arbitrageurs looking to seize unexplored opportunities and seek out new and better ways of doing things.[13] In *Innovation and Its Enemies*, Calestous Juma documents numerous historical case studies of innovators bucking traditions and laws to offer the public new goods and services, most of which would later come to be considered vitally important and socially beneficial inventions. Those innovations included things like coffee, margarine, the printing press, farm equipment, electricity, mechanical refrigeration, recorded music, transgenic crops, and genetically engineered salmon.[14] In each case, "tensions between innovation and incumbency" led to social skirmishes and legal battles about the boundaries of acceptable practice.[15] Some people had to put their necks on the line and dare to be different to advance a better idea or introduce a better—or entirely different—way of doing things. In the process, those innovators were often acting in an evasive (and potentially even illegal) fashion.

Making Permissionless Innovation a Reality

This cycle seems to be accelerating today, however. Evasive entrepreneurs are increasingly pushing permissionless innovation as a policy prerogative and often follow the maxim known as Hopper's Law, which is named for Grace M. Hopper, a computer scientist who served as a rear admiral in the U.S. Navy. Hopper once famously noted, "It's easier to ask forgiveness than it is to get permission."[16] The phrase "permissionless innovation" is sometimes associated with Hopper's Law, although it does not appear Hopper herself ever used that specific phrase.[17]

Following the logic of Hopper's Law, permissionless innovation caught on in computing engineering circles and then more generally in high-tech sectors after the success of the internet and Silicon Valley companies throughout the 1990s and beyond. Permissionless innovation came to represent the defining zeitgeist of the internet era and an operational ideal to which a great many information technology creators and companies aspired. It represented the general belief that there was great benefit in letting people try new things without asking for the blessing of any number of unnamed authorities or overseers, which could include businesses, bosses, and perhaps even government officials.

As I noted earlier, my previous book on permissionless innovation made the case for expanding this vision and making it the presumption underlying technological governance across many sectors, beyond just the information technology sector. Generally speaking, experimentation with new technologies and business models should be permitted by default as a matter of public policy, except in those cases where clear and immediate harm is evident.[18] The connection between permissionless innovation and evasive entrepreneurialism should be obvious. Evasive entrepreneurs seek to make permissionless innovation their operational lodestar in a business sense and sometimes as a policy prerogative too.

But my earlier book also argued that permissionless innovation must not be equated with anarchy. There will always be a need for rules to govern markets and innovative activities. As a policy vision, permissionless innovation represents the belief that restrictions on human creativity should generally be our last resort, not our first. Innovators should be innocent until proven guilty. Generally speaking, therefore, ex post remedies are preferred to ex ante restraints on creative activities. Yet, as later chapters will explore in greater detail, anticipatory governance efforts are sometimes needed to address more serious risks before they unfold. Thus, even though I will argue that evasive entrepreneurialism can have many benefits for society, there will be times when it must be restrained in various ways to ensure that other values are protected. As we will see, however, striking that balance is rarely easy as a matter of either policy balancing or practical enforcement efforts.

The Evasive Entrepreneurialism Playbook

We are now in a position to think about the playbook that entrepreneurs sometimes use when they seek to act in an evasive fashion or to become regulatory entrepreneurs.

Elizabeth Pollman and Jordan M. Barry note how today's regulatory entrepreneurs typically "seek to grow 'too big to ban' before regulators can act."[19] They "rally the public to their cause, and then use their popular support as leverage to win the change they want from resistant officials."[20] The tools they are creating can help make it easier for them to mobilize their user base to become citizen lobbyists on their behalf and rally users around the idea of making permissionless innovation the default public policy approach to new technologies.[21]

Uber, the ridesharing company founded in 2009, is a prominent example of a regulatory entrepreneur that has employed these tactics and used "civil disobedience as a business model."[22] The Uber case study is important because it serves both as an instructive story about the potential for evasive entrepreneurialism to change laws and markets and as a cautionary tale about the limits of evasiveness and permissionless innovation in practice.

When it first began expanding its service, Uber aggressively entered local transportation markets across the globe without first seeking formal permission from most regulatory authorities. Instead of caving to political demands, the firm tapped the power of its network of drivers and customers to lobby on its behalf.[23]

The crucial feature of Uber's approach to changing policy was the way it calibrated evasive activities to achieve greater leverage in political negotiations, not necessarily to pretend that it could forever ignore politics and regulation altogether. Correspondingly, this method may be a way to help minimize regulatory costs at some later point in time.[24] It is a core feature of regulatory entrepreneurialism in many other contexts, as we will see in other examples discussed later.

Uber's experience in New York City represents a textbook example of regulatory entrepreneurialism and is worth considering at greater length. Following an effort by New York mayor Bill de Blasio to push legislation to limit the expansion of ridesharing in the city, the company quickly "mobilized our customers, over 100,000 of them, either e-mailed or tweeted at City Hall or the city council," according to Bradley Tusk, one of Uber's political strategists.[25]

Uber even went so far as to recode its mobile phone app in New York City such that when users scrolled through their transport options, one choice read "DE BLASIO."[26] Users who clicked on it were shown that either no Uber drivers were available or they should expect wait times of 25 minutes for a ride. It also included a note to users calling on them to email the mayor's office and the city council to stop the legislation.[27] Uber's intention was to "demonstrate what life for NYC riders would be like if de Blasio's plan to limit Uber [were] passed into law."[28] *Observer* reporters Jillian Jorgensen and Will Bredderman describe the pressure that followed:

> The mayor was pummeled by the apps, particularly Uber, which ran ads highlighting the service's popularity with lower-income, minority New Yorkers who were trying to earn some extra money as drivers or were just happy to

finally get a cab in a part of town where yellow taxis wouldn't venture. In-app notifications allowed Uber users to e-mail the mayor's office with their displeasure. Gov. Andrew Cuomo waded into the fight to say the state and not Mr. de Blasio should regulate Uber. Council members were slammed by direct mail touching on the same topics as the television commercials.[29]

In the short run, Uber's evasive efforts worked. The pressure the company, its drivers, its users, and the public put on the city forced de Blasio and advocates of ridesharing regulation to back off.[30] Uber successfully used the too-big-to-ban strategy, making it difficult for the city to clamp down on its service.[31]

Uber has been able to use this approach so frequently and effectively—both domestically and globally—that some have come to call it Travis's Law, named after former Uber CEO Travis Kalanick. Journalist Brad Stone describes this "law" as follows:

Our product is so superior to the status quo that if we give people the opportunity to see it or try it, in any place in the world where government has to be at least somewhat responsive to the people, they will demand it and defend its right to exist.[32]

To be sure, that somewhat arrogant stance will antagonize policymakers, incumbent operators, and other defenders of the regulatory status quo. After all, there is a process, and people are expected to follow it. The problem is that the process that ridesharing companies like Uber, Lyft, and other transportation innovators were expected to adhere to did not leave much room for new competitors. That process was fundamentally broken and generally working at odds with public welfare because of cronyism and regulatory capture by incumbent operators.[33]

For many decades, economists and scientists had documented the inefficient and anti-consumer way that local transportation regulatory schemes function.[34] Powerful special interests had captured the regulatory process and used it to their advantage, keeping competition out and ensuring prices remained high.[35] The Federal Trade Commission had even opened investigations into these schemes and concluded "that no persuasive economic rationale is available" for protectionist and anticompetitive taxicab regulations and that licensing restrictions "waste resources and impose a disproportionate burden on low income people."[36]

The sort of chronic cronyism and regulatory capture at work in that market is a feature of many other sectors of the economy, as will be noted in a subsequent chapter.[37] For now it is enough to note that despite widespread consensus that local taxi and transportation licensing schemes were a nightmare of cronyism that hurt competition and consumer welfare, nothing much ever happened to address these problems. In fact, things seemed to get worse year after year. And then Uber, Lyft, and other ridesharing innovators were launched and the public was finally given a taste of true competition and choice. The public liked it so much that these firms, *and their drivers and users*, were largely able to engage in an end-run around the regulatory state.

To a slightly lesser extent, Airbnb followed the same model of regulatory entrepreneurialism. They and other rivals to the hotel and lodging industry were able to buck traditional regulations and gain enough leverage to negotiate more reasonable regulatory treatment over time. To be clear, this treatment by regulators did not mean the complete abandonment of all local lodging rules. In fact, Airbnb and other space-sharing companies were eventually forced to abide by some of the local rules and collect some lodging taxes, just like traditional hotels.

But what is important here is that their evasiveness prompted a more reasonable negotiation with local regulators as competition slowly took root.[38] In most cases, the incumbent operators were not able to drive these new rivals out of town to protect their old business models. Consumers were finally treated to new choices precisely because the system *did not* work, and innovators decided to become regulatory entrepreneurs and work to change things for the better, albeit in an unconventional and often controversial fashion.

Can the Uber and Airbnb stories serve as a blueprint for other evasive entrepreneurs to follow? One serious problem with that story of change is that it involves Uber, which has become a corporate pariah of sorts in the eyes of many critics who decry the firm's "reputation for lawlessness" and "toxic culture of rule breaking."[39] Part of this opposition to Uber's approach arose because of the behavior of Kalanick, who eventually resigned after various controversies and public embarrassments. Despite the contempt many hold for Kalanick, it is still worth considering whether pushing the envelope aggressively as Uber did can serve as a successful business strategy in other contexts.

Some cities, like Austin, Texas, did call Uber's bluff and pushed back so forcefully that Uber pulled out rather than comply with city ordinances. Following a dramatic exit, it returned to Austin after political compromises were negotiated.[40] Even in New York City, the site of its greatest victory, Uber and fellow ridesharing companies continue to feel regulatory pressure in the form of new limits on how many ridesharing vehicles can operate in the city.[41] In late summer 2018, the New York City Council passed legislation limiting new licenses of for-hire vehicles like Uber and Lyft.[42] This legislation shows how, with enough pressure, lawmakers and regulators will often still be able to make innovators cave to certain political demands.

Meanwhile overseas, in 2017, London withdrew a license Uber needed to operate in the city, citing concerns that the firm had shown a "lack of corporate responsibility."[43] Uber never stopped operating in London while the dispute was in court and recently came to a modified 15-month agreement with the city that will make Uber more beholden to the city's transportation regulators.[44] Uber has opted to bend the knee rather than risk permanent exile from one of its most important markets.

Still, what Uber and other ridesharing companies accomplished with their evasive strategies both here and abroad seems undeniable. The company was operating in more than 70 countries by the end of 2018, and it has inspired a host of copycat companies, not only in the transportation sector but also in many other fields, as other would-be disruptors seek to be the "Uber" of their respective industries.[45] This playbook seems to have worked for Airbnb and other space-sharing platforms, too. By putting consistent pressure on traditional regulatory regimes across the world, these innovators were able to open up markets that were closed to competition and consumer choice for many years. That same blueprint could serve as a model for other innovators who are willing to become both marketplace entrepreneurs and regulatory entrepreneurs. Rentable electric scooter providers seem to be using a similar playbook, as will be examined later.

Recent books by Evan Burfield (*Regulatory Hacking*) and Bradley Tusk (*The Fixer*) build on their experiences as a startup investor and a consultant, respectively.[46] Burfield and Tusk guided many disruptive innovators through political minefields and helped them to successfully use evasive techniques and a permissionless-innovation approach to disrupt the status quo and achieve greater operational freedom for those firms.[47] Both authors identify common strategies that evasive entrepreneurs can use, including mobilizing existing or potential customers to assist as citizen lobbyists. They also note the importance of good marketing and public relations when it comes to framing the debate. Their experience suggests that there is a set of strategies other would-be evasive entrepreneurs could tap to launch new ventures, even if the political and regulatory world is not yet ready for them.

However, there will be other cases in which the evasive entrepreneurialism or regulatory hacking playbook will backfire, and innovators must be prepared for that danger. The most notable recent example involved home genetic testing innovator 23andMe. Originally, 23andMe offered a direct-to-consumer test kit that let users screen for roughly 250 genetic conditions, but the company did so without seeking the blessing of regulators at the Food and Drug Administration (FDA).[48] The FDA eventually responded aggressively and ordered 23andMe to stop selling its kits and then come back to the agency and ask for permission to remarket a more limited home genetic test.[49] Instead of fighting the order in court, the company negotiated with the FDA to get a new test on the market, which analyzed just 10 conditions.

The 23andMe case study serves as an example of how difficult it will be for evasive tactics to work in sectors that are born in what we might think of as "regulatory captivity" instead of being born free. I will return to that distinction later in this chapter and consider what it means for other technologies and the prospects of regulatory entrepreneurialism. But first it is important to examine how innovation arbitrage plays into this story.

Innovation Arbitrage

Evasive entrepreneurialism can also involve a certain amount of jurisdictional shopping, or innovation arbitrage. This is the idea that innovators can move their operations to jurisdictions that provide legal and regulatory environments that are more hospitable to entrepreneurial activity.[50] Just as capital now fluidly moves around the globe seeking more friendly regulatory treatment, the same is increasingly true for innovations. This situation also plays out domestically when innovators seek to play state and local governments off each other in search of some sort of competitive advantage.[51]

Such arbitraging opportunities are accelerating for individuals as well. "Brain drain" is a term that gained currency in the 1960s and 1970s to describe the movement of talented scientists and technologies from one jurisdiction to another.[52] This process is crucial today more than ever before. In essence, the same globalization trends that have made it easier for goods, services, and capital to be produced and used anywhere in the world are now also a driving force in the information technology realm and many other technology sectors.[53] Back in 1993, before the internet had even gone mainstream, Nobel Prize–winning economist Milton Friedman was predicting how new information technologies would make innovation arbitrage easier. "The technological revolution has come from developments in the area of computers and telecommunications, and this has made it possible, to a far greater extent than any time in the world's history, for a company to locate anywhere, to use resources from anywhere to produce a product that can be sold anywhere," he said.[54]

Companies are doing that today in many sectors beyond information technology because, as Jason Potts writes, "entrepreneurship is inherently global in orientation. It does not naturally stop at the boundaries of a nation state."[55] In his 2016 book, *The Great Convergence*, Richard Baldwin identified how information technology has spurred a "New Globalization," which has "denationalized comparative advantage by redrawing the international boundaries of competitiveness," meaning that "national boundaries are no longer the only relevant frontiers when thinking about international competition."[56] In this more globalized and hyperconnected world of fragmented production, burdensome rules that prohibit innovative activities can encourage firms and individuals to offshore their operations to jurisdictions with less onerous regulations.[57]

Innovators can now take advantage of the fact that modern technology, as James Dale Davidson and William Rees-Mogg write, "divorces income-earning potential from residence in any specific geographic location."[58] "If rules differ across polities (cities, states, countries), an entrepreneur can exploit these institutional inconsistencies by locating where rules are less binding or less enforced, provided that there is free movement," notes administrative law expert Alfred Aman.[59] "As internationalization progresses, such cross-border institutional arbitrage is becoming increasingly important."[60] The lure of China's massive marketplace and growing technological capabilities could prove particularly tempting for many innovators and investors in this regard.[61] But many other major tech hubs already exist across the globe that are luring technology entrepreneurs away from Silicon Valley and the United States more generally.[62]

One good example of innovation arbitrage in practice involves commercial unmanned aircraft systems (UAS), or drones, as they are more commonly called. In late 2013, American tech giant Amazon decided to begin research into the possibility of delivering its packages using drones. The experiments were so constrained by Federal Aviation Administration (FAA) regulations that the firm opted to move much of its research and development, as well as operational testing, to the United Kingdom and Canada.[63] Until late 2016, the FAA had no formal drone regulations and effectively banned all commercial drone activity. Unsurprisingly, other U.S.-based drone innovators decided to move abroad to develop their products and services without re-strictive regulatory constraints.[64]

Even though the FAA has since established rules for commer-cial and recreational drones, many experts still lambaste the FAA for taking an extremely risk-averse approach to airspace in-novation.[65] A strongly worded 2018 report by the National Acad-emies of Sciences, Engineering, and Medicine found that "'fear of making a mistake' drives a risk culture at the FAA that is too often overly conservative, particularly with regard to UAS tech-nologies."[66] Because "the status quo is seen as safe" and the FAA has "an overly conservative risk culture," the report concluded that the agency's policies have created "a significant barrier to in-troduction and development of these technologies," which was hurting public safety by discouraging life-enriching and even life-saving innovations.[67] In the FAA Reauthorization Bill of 2018, Congress sided with the National Academies report, order-ing the FAA to take a reasonable, risk-based approach to drone regulation.[68]

The FAA's risk-averse culture is changing, but it may be too little, too late. The FAA's slow-walking of liberalizing drone use in the United States has given other nations the first-mover ad-vantage in services like drone delivery.[69] By the time U.S. regula-tors finally approved Amazon's initial application for drone delivery services, the firm was no longer operating the particular prototype for which it had originally applied for permission.[70] It is another example of technological developments moving faster than political developments, that is, the pacing problem.

The consequences of regulatory delay are important for more than just the convenience that drones might provide, like delivery services. Drones have many life-saving applications, too, as they have already been used to monitor or fight forest fires,[71] help with search-and-rescue missions[72] for missing people or animals,[73] assist lifeguards by dropping life preservers to drowning people,[74] deliver medicines to remote areas,[75] and more.[76] Policies are slowly starting to change in the United States to allow more innovations like those to occur.[77] In May 2018, several state, local, and tribal governments were selected to serve as sandboxes for drone testing, with the hope that the results will encourage the FAA to lift restrictions on nighttime and overhead flights as well as flights beyond the operator's visual line of sight.[78]

Innovation arbitrage seems to also be taking hold in genetic testing and editing. When the FDA ordered 23andMe to stop marketing its at-home genetic analysis kit in 2014,[79] the company was quickly courted by government officials in the United Kingdom.[80] The U.K.'s Medicines and Healthcare Products Regulatory Agency said 23andMe's test could be used there with some loose limits.[81] 23andMe was also simultaneously making its Personal Genome Kit available in Canada, Denmark, Finland, Sweden, and other countries. It is worth noting that there were no reports of adverse effects to consumers in those countries because of the availability of 23andMe's test, which suggests that the FDA's fears and efforts to pull the test off the market in the United States may have been unnecessary. Regardless, what is important here is that 23andMe did not stop innovating simply because its primary domestic regulator moved to restrict its service. The technology and the company moved forward. This sort of activity will be increasingly likely in a world where innovation arbitrage opportunities multiply. Some experts suggest that the United States and Europe are already "losing the gene editing race" to China and other Asian countries, which are investing massively in these technologies and also offering a more flexible regulatory environment.[82]

Innovation arbitrage happens within national borders, too, especially in countries with a federal system of government like the United States. Throughout the United States, state lawmakers attempt routinely to lure innovators away from less favorable regulatory environments in the hope of attracting a better job-creating and tax-generating economic base to their own jurisdictions.[83] Some of those incentives for innovators arrive in the form of tax inducements, which often backfire and cost governments and taxpayers more than they are worth.[84] But local governments are also tailoring public policies in such a way as to make their jurisdictions more hospitable to innovators.

With the growth of autonomous vehicles, for example, states like Arizona,[85] Florida,[86] and Ohio[87] have moved quickly to make it known that they would provide a more hospitable regulatory environment for autonomous cars and trucks than more restrictive states like California. As a result of this sort of competitive federalism, more restrictive states have attempted to modify such regulations after the fact to reattract innovators and technology.[88]

San Francisco was one city on the losing end of this process when, in late 2016, Uber famously packed up all its driverless cars and moved them to Arizona to do testing there.[89] The move followed an effort by Arizona officials to court the firm to come to their state, which included a statement from Governor Doug Ducey that "Arizona welcomes Uber self-driving cars with open arms and wide open roads."[90] Later, however, an autonomous Uber car struck and killed a pedestrian in Tempe, Arizona, leading the company to take its vehicles off public roads for a time.[91] Nine months later, however, Uber resumed testing in another state after the company was cleared of wrongdoing in the Arizona crash.[92]

Similarly, San Francisco's restrictive approach to delivery robots pushed innovators in that space to relocate. Companies like Marble and Starship Technologies had been testing their self-driving delivery robots in the city beginning in 2016. These small robots, which look like coolers on wheels, were able to bring food, groceries, and other deliveries directly to a customer's door. But in late 2017, San Francisco imposed extreme restrictions on delivery robots, citing fears about them overcrowding sidewalks or affecting delivery jobs done by humans.[93] City officials were even considering outright bans on delivery robots or heavy taxes on them to counter the feared impact on jobs, although they settled on highly restrictive rules for how and where the robots could be operated. That was enough to encourage some companies to leave town. Ahti Heinla, CEO of Starship Technologies, told *Business Insider*, "[T]here are many more cities that are welcoming our robots and that want to work with us," so his firm decided to leave San Francisco to focus on markets where such innovations would be more welcome.[94] Artificial intelligence (AI), robotics, and biotechnology are fields in which global innovation arbitrage could become more prevalent, with innovators in China, Canada, and many European countries vying with America for the early lead in these fields.[95]

Individuals also engage in a form of innovation arbitrage. Medical tourism—going offshore to seek out alternative medical treatments not available domestically—is nothing new, but it is getting easier, thanks to technological advances and cheaper travel options. In one particularly notable example, in 2011, NBA basketball superstar Kobe Bryant traveled to Germany to receive "platelet rich plasma" therapy for his arthritic knee.[96] He engaged in that medical tourism because biological treatments face stricter regulation by the FDA in the United States and "most American doctors are unwilling to risk the ire of regulators."[97] Medical tourism can involve controversial practices and thorny ethical dilemmas, many of which deserve regulatory oversight.

But that doesn't mean regulation will be easy, and in some cases it is not clear who would be conducting such oversight.

For example, in fertility clinics in Ukraine, doctors are already helping infertile patients get pregnant using the DNA from three different people.[98] Such "three-parent baby" procedures are banned by the FDA, but what should U.S. officials do about Americans traveling abroad in search of these treatments? Meanwhile in China, genetic editing is advancing so rapidly that in late 2018, a medical team at the Shenzhen-based Southern University of Science and Technology began recruiting couples to create gene-edited babies.[99] Their goals were to edit the genetic code of unborn children to make them resistant to HIV, smallpox, and cholera.[100]

We should expect innovation arbitrage—by those both developing and benefiting from innovations—to become increasingly prevalent in the field of medicine precisely because so much is on the line. Regulation will likely require international treaties and multinational enforcement efforts, but such efforts will prove challenging as travel costs fall, enforcement costs rise, and ease of access to new technological capabilities continues to expand.

How extreme can global innovation arbitrage get? Consider rogue satellite launches. In April 2018, a Silicon Valley startup called Swarm Technologies launched four small satellites into space from India as part of a new venture to create a space-based Internet of Things communications network.[101] However, Swarm had not yet received permission from U.S. regulators at the Federal Communications Commission (FCC) to launch its constellation of satellites and provide wireless service to American citizens from space.[102] It will be interesting to see if Swarm's combination of technological civil disobedience and global innovation arbitrage works in this case or instead results in a regulatory backlash. Although Swarm is a long way from its ultimate goal of gaining operational freedom to serve the public, its four microsatellites have been picked up by the U.S. satellite tracking system, which the FCC believed would not be the case when the agency denied launch authorization.[103] The FCC granted temporary permission to Swarm to transmit orbital and tracking data, which suggests that the firm's gambit still could work.[104]

Innovation arbitrage will not always be an all-or-nothing affair. As we will see, the threat of offshoring innovation will sometimes merely be a bargaining tactic used as part of a broader evasive entrepreneurial strategy. Other times, entrepreneurs might move some but not all their operations, or move some of them only to move them back later after policies have changed. For example, drone innovators did push many of their testing operations offshore in past years, but they also continued to push for reforms here in the United States while continuing research abroad. The strategy appears to have paid off as Congress and U.S. regulators recently endorsed and established trials for drone deliveries[105] and cleared a path for commercial drone deliveries to be legalized.[106]

This example reflects the finding discussed earlier that evasive entrepreneurs are not always out to evade laws or regulations entirely but rather to create more leverage in negotiations with policymakers to achieve a superior result once regulation of some sort becomes inevitable. Although many officials and critics may lambaste entrepreneurs for using the threat of innovation arbitrage, it will likely remain a powerful bargaining tactic in years to come as the pacing problem accelerates.

Finally, the growth of global innovation arbitrage has ramifications for countries and their competitiveness, as well as their ability to attract and retain human capital. To some extent, countries will have to adapt their policies and cultures to accommodate a great deal more entrepreneurialism, or else they could lose out to others in the race for global competitive advantage. "Cultures that attempt to block technology for reasons that appear desirable will, all things equal, eventually be dominated by those that embrace it," says Braden Allenby, an environmental scientist at Arizona State University. He notes that this "obviously poses an unhappy dilemma" to many countries that do not wish to adapt to these realities because, "in a highly competitive global environment, where many cultures are jostling for position, technological evolution will be difficult, if not impossible, to stop."[107]

The United States is not immune to these pressures despite having a lead in many high-tech sectors. If policymakers erect more obstacles to innovation, it will encourage entrepreneurs to look elsewhere. For example, in late 2018, the U.S. Department of Commerce's Bureau of Industry and Security announced a "Review of Controls for Certain Emerging Technologies," which launched an inquiry about whether to greatly expand the list of technologies that would be subject to America's complex export control regulations.[108] Most of the long list of technologies under consideration (such as artificial intelligence, robotics, 3D printing, and advanced computing technologies) were dual-use in nature, meaning that they have many peaceful applications in addition to military ones. If restrictive export controls were imposed on such technologies, it would likely undermine U.S. innovation and competitiveness.[109] Commenting on the effect such rules might have, the *New York Times* suggested that "[o]verly restrictive rules that prevent foreign nationals from working on certain technologies in the United States could also push researchers and companies into other countries."[110] The *Times* also quoted an international trade lawyer who said of the rules, "It might be easier for people to just do this stuff in Europe," if the

United States imposed controls.[111] Anti-innovation policies create incentives for entrepreneurs to behave even more evasively.

Other Evasive Entrepreneurialism Case Studies

With this framework for understanding evasive entrepreneurialism in mind, we will explore some case studies of it in action across many technological sectors. In doing so, we will consider some questions about the ethical and legal dimensions of each example to help frame the discussion that ensues throughout the rest of this book.

MOBILE MEDICAL APPLICATIONS

You may think of that device you carry in your pocket as a smartphone, but increasingly it is a mobile medical device as well. Powerful new capabilities are already built directly into smartphones and smartwatches that enable step-counting, heart rate measurement, and more. More sophisticated applications allow users to track diet and weight patterns, address vision deficiencies, correct sleep disorders, or take pictures of their skin and connect directly with dermatologists for advice about potential treatment for abnormalities. These technologies are growing rapidly as consumer demand intensifies for more personalized medicine and tailored health and fitness options through wearable technology. Much of this demand is driven by movements called "quantified self" and "wellness tech." They include a diverse array of tech-enabled efforts to better track and address fitness and wellness.[112] Eric Topol, MD, calls those movements "the creative destruction of medicine" and predicts that "in the coming years, we'll see [digital applications] for measuring blood glucose, sleep brain waves, and all vital signs, stress, and mood quantified. Measuring vitals will eventually be as common as counting calories or the number of steps you've walked."[113]

Many of the mobile medical applications now available in smartphone app stores (as well as other quantified self technologies) would have likely been illegal in years past because they did not go through the FDA's formal medical-device approval process. Yet, with the rapid pace of change—with health and fitness apps and devices multiplying faster than regulators' ability to monitor them—the FDA has been forced to rethink its approach to these tools, which it calls "digital health technologies." The agency has largely given up trying to police low-risk applications, like fitness trackers, online medical dictionaries, and other health information services.[114] The agency has tried instead to focus on potentially more dangerous digital health technologies, such as those designed to treat or diagnose medical conditions.

The FDA is trying to formalize this new approach as part of its Digital Health Innovation Action Plan, which was announced in mid-2017.[115] The goal, the agency said, was "to reimagine FDA's approach for assuring that all Americans, including patients, consumers and other health care customers have timely access to high-quality, safe and effective digital health products."[116] That plan included a program called Pre-Cert (short for "pre-certification") aimed at creating "a future regulatory model that will provide more streamlined and efficient regulatory oversight of software-based medical devices developed by manufacturers who have demonstrated a robust culture of quality and organizational excellence, and who are committed to monitoring real-world performance of their products once they reach the U.S. market."[117] In late 2018, the agency also announced a new De Novo Classification Proposed Rule that would clarify agency rules aimed at approving new low-risk or moderate-risk medical devices.[118]

These recent FDA efforts blend traditional elements of hard law (formal legislative statutes and agency regulations) with soft law, which represents less formal and more adaptable governance measures than traditional rules require. Such soft-law efforts incorporate a broad range of more flexible governance options for emerging technologies, including agency guidance documents and consultations, multistakeholder processes, ongoing consultations, the formation of best practices, and education efforts. For now, it is enough to note that as wearable tech, the quantified self movement, and the idea of "software as a medical device" take hold,[119] more flexible soft-law governance mechanisms will likely need to supplement or potentially even replace many traditional regulatory schemes.[120]

As we will see with some of the examples discussed next, many other technologies are being used in conjunction with computers, smartphones, and online services to create innovations that sometimes push the boundaries of legality. What is clear, as my former colleague Jordan Reimschisel recently observed, is that "technology will usher in a world where medicine is democratized and decentralized whether those in Congress and at the FDA like it or not."[121] Citizens demand more and better health and wellness technologies, and innovators are responding, often by acting as evasive entrepreneurs. Policymakers must modify traditional policies quickly or else run the risk of being made less relevant through evasive entrepreneurialism in the health care arena.

3D PRINTING

Some citizens are using 3D printers to collaborate on projects, such as the creation of prosthetic limbs[122] or weapons[123] or even replicas of popular toys.[124] Although these collaborations are voluntary and usually noncommercial, they potentially violate various food and drug regulations, firearms registration mandates, and intellectual property laws, among others.

The most interesting example of this phenomenon involves the volunteer effort called "e-NABLE," which is short for "Enabling the Future." The volunteer organization collaboratively designs and creates 3D-printed prosthetics for individuals (especially children) with limb deficiencies.[125] These volunteers share open-source blueprints and other information on various websites with others across the world.

Meanwhile, others are creating custom 3D-printed orthoses to help children with cerebral palsy walk comfortably and without the aid of crutches. Because off-the-shelf solutions are often ineffective and uncomfortable for many kids, some parents have custom-made orthoses for their own children to help them successfully walk.[126]

Amateur prostheses are also being widely distributed today and helping to save many individuals and families a significant amount of money. But prostheses are medical devices in a traditional regulatory sense, and no one making his or her own is going to the FDA to ask permission for, or a right to try, new 3D-printed limbs.[127] Instead, that person will just go ahead and make new prostheses for people in need.[128] How should we regulate all this bottom-up innovation by average citizens (especially considering how much of it is noncommercial in character)?

Reason magazine contributing editor J. D. Tuccille argues that, thanks to the growth of additive manufacturing, "you'll soon be able to manufacture anything you want and governments will be powerless to stop it."[129] That prediction is not entirely likely, however, because governments may still be able to regulate the outputs of 3D printers.

Consider 3D-printed firearms. Distributed Defense has challenged existing gun regulations by freely distributing blueprints to fabricate plastic guns at home. There is not a large market for such guns currently, but the very threat of such decentralized weapon creation has led to lawsuits by government officials who hope to somehow repress the production and dissemination of 3D gun blueprints over the internet.[130]

Those open-source blueprints represent free speech, however, which means that the First Amendment likely protects creation and dissemination.[131] Moreover, such blueprints have been available on various file-sharing sites for many years, making it hard to put the genie back in the bottle in this case.[132]

Public officials and other gun control advocates, however, remain worried about unregulated access to do-it-yourself (DIY) firearms, or "ghost guns," even though such weapons have not been shown to be a major problem.[133] But it is only going to get easier to make stuff with these new technological capabilities, so enforcement concerns will likely intensify.[134] As DIY weapons become more prevalent, it is worth remembering that a gun is a gun regardless of how it is manufactured, and governments will still regulate how and where guns are used. The same is true for 3D-printed medical devices or other products. Decentralized production of those things will make point-of-production enforcement significantly more challenging and costly. For better or worse, acts of evasive entrepreneurialism and technological disobedience will become more likely as 3D manufacturing capabilities spread and become more popular.

DIY MEDICAL DEVICES AND BIOHACKING

DIY health services and medical devices are on the rise thanks to the combined power of open-source software, 3D printers, cloud computing, and digital platforms that allow information sharing between individuals with specific health needs. Average citizens are using these new technologies to modify their bodies and abilities, often beyond the confines of the law.

Welcome to the occasionally scary but oftentimes awe-inspiring world of biohacking. Biohackers are essentially "pro-sumers," the term many used a decade ago to describe the way average citizens were taking advantage of new communications and computing technologies to become both producers and consumers of news, information, and entertainment.[135] Pro-sumers evaded traditional industry norms and government regulations that had previously made it difficult for citizens to communicate freely. The same phenomenon is now shaking up the world of health and medicine as pro-sumers use new technological capabilities to take their health into their own hands and likely evade many traditional norms and regulations when doing so.

Some biohackers understand their role to be that of a "community biologist, do-it-yourself biologist, or even a citizen scientist" who conducts "open-source science."[136] Others refer to this movement as "patient-led research" or "citizen-driven biomedical research."[137] Regardless of what it's called, it appears to be growing and creating headaches for regulators as it does. As discussed earlier, diverse parties are already coming together and using open-source software and 3D printers to fabricate prosthetic devices for children and other individuals with limb deficiencies. Two similar examples involve insulin pumps and teeth aligners.

The Nightscout Project is a nonprofit founded by parents of diabetic children.[138] These parents came together and shared knowledge and open-source code to create DIY insulin remote monitoring and delivery devices for their kids.[139] Nightscout's motto is "WeAreNotWaiting." Specifically, these parents got tired of waiting for new professional devices to be developed and then approved by the FDA. It can take many years for such devices to get through the regulatory approval process. Through voluntary collaboration, these parents have created reliable devices that are much less expensive than the FDA-approved devices, which can cost many thousands of dollars.[140] Sonya Collins of *Genome* describes the movement as "an international community of patients and care-givers who have hacked, crowd-sourced, and data-shared their way to the medical technology they want rather than wait for the [FDA] to approve a new device."[141]

Should the parents who make such devices be fined by the FDA or penalized in some fashion? Or should they be regulated through a licensing process—thus essentially disallowing them from using technologies to freely treat their own children at a lower cost? Regulators are already struggling with such questions. Consider the story of one of these DIYers. Bryan Mazlish created a successful artificial pancreas system for his wife and son, who both have Type 1 diabetes, and he hoped to share his design with others. Collins notes that when Mazlish ran the idea past regulators, they "took the wind out of his sails" after informing him that "even if he wanted to give it away for free on the internet, he'd still have to get FDA approval."[142] But Mazlish and other Nightscout volunteers lacked the resources to pursue FDA approval, which takes years and can be quite costly. They even considered giving the design away to existing device manufacturers but could not find any takers. The uncertainty surrounding the product's legal standing with regulators likely limited Nightscout's ability to expand their efforts. Mazlish's disappointment was palpable. "I can't tell you how hard it is to have such a life-transforming system and have to tell someone else who could benefit from it, 'We can't give it to you,'" he said.[143]

The FDA's precautionary approach is based on a risk-averse culture that seeks to protect the health and safety of Americans by making sure dangerous devices don't get to market. But in this case, it seems obvious that the approach is having serious unintended consequences by stifling bottom-up innovation that could actually improve (or even save) lives and at lower cost.

But these restrictions haven't stopped other citizen innovators, and, as witnessed with the rise of Nightscout, this movement is likely to expand in coming years as these technological capabilities become more available. "High Insulin Prices Drive Diabetics to Take Extreme Measures," a late 2018 *Wall Street Journal* headline noted.[144] Fueled by social media sites and other online platforms, diabetics are connecting with each other and finding ways to create a gray market in alternative DIY medicines. *Bloomberg Businessweek* reporters Naomi Kresge and Michelle Cortez investigated the growth of this market and found that:

> By some estimates, as many as 2,000 people around the world have used a home-built pancreas, cobbled together mostly via social media and the free-code clearinghouse GitHub. Tech support consists of parents and patients who use Facebook Messenger or email to help newcomers fix bugs or revive busted equipment. There are plenty of potential converts: In the U.S. alone, about 1.3 million people have Type 1 diabetes, and there are indications the technology could also help some sufferers of Type 2, the group that accounts for most of the world's 422 million diabetes cases.[145]

Where else might such bottom-up, free innovation take root and challenge traditional rules and regulations? Consider the case of do-it-yourself dental aligners. In 2016, Amos Dudley, a 23-year-old college student with no previous dentistry experience, was able to use a 3D printer and a laser scanner at his university to make his own orthodontics for just $60.[146] That price represents significant savings off commercial aligners, which can run thousands of dollars. Dudley's DIY plastic braces helped align his teeth at great savings while also "stick[ing] it to the dental appliance industry."[147] However, it was also a dangerous experiment that could have put him, and others who followed his lead, at risk.[148] But what should the law say about people like Dudley, who are creating their own specialized medical devices in an open-source, noncommercial fashion? Although he did not do so, what would regulators have done if he had helped others create similar devices, whether free or for a small fee?

What does the future hold for such biohacking activities? One possibility is that such technologies go mainstream and become more commercialized. Dental aligners provide an example of this. What Amos Dudley did was never going to scale. Most people have no desire or ability to fabricate their own dental aligners. But others can fill the void. Small startups like SmileDirectClub are challenging the more established invisible braces industry by offering cheaper teeth straightening without all the office visits to licensed dentists.[149] Unsurprisingly, professional dentistry organizations such as the American Dental Association and the American Association of Orthodontists oppose such technologies on safety grounds, although it is also obvious that they probably do not want the added competition.[150]

Another future scenario is that biohacking remains decentralized but becomes more mainstream and professional without becoming fully commercial. Many biohackers already come together to pool knowledge through community labs or "biohackerspaces."[151] These "community labs are popping up to bring hands-on learning opportunities to amateur scientists desiring to learn more about the secrets of DNA."[152] Biohackerspaces can be found in several U.S. cities today, including Brooklyn (Genspace); Baltimore (Baltimore Underground Science Space); Berkeley, CA (Berkeley BioLabs); Norfolk, VA (Biologik); and others.[153] The growth of biohackerspaces is reminiscent of the rise of community *maker spaces*, where average citizens come together to share knowledge and use tools they probably do not own (like 3D printers and laser etchers) to create interesting new things. The difference is that these biohackerspaces are legitimate laboratories with scientists on hand to help people learn more about new genetic capabilities.

Another future scenario would see biohacking turn even more rogue or underground in nature as a form of guerrilla innovation that sometimes borders on neo-anarchism. A good example of this involves Michael Laufer and the Four Thieves Vinegar group. The group's name is apparently borrowed from an elixir thought to protect against the plague in past centuries. As the folk story goes, thieves who robbed those dying during a European plague were able to pull off their heists without getting sick because they had discovered a recipe combining vinegar and various herbs (including garlic, rosemary, and sage) that immunized them from disease. When they were later apprehended, they surrendered their secret formula in exchange for an easier death sentence.[154] Although likely apocryphal, the story continues to inspire others to experiment with similar formulas in an attempt to ward off various diseases.[155]

Laufer and his fellow Four Thieves Vinegar biohackers have been described as "a volunteer network of anarchists and hackers developing DIY medical technologies."[156] The group's motto is "Free Medicine for Everyone," and they work to provide it by intentionally evading various legal and business obstacles. "To circumvent these, we have developed a way for individuals to manufacture their own medications," the group's website says.[157] "We have designed an open-source automated lab reactor, which can be built with off-the-shelf parts, and can be set to synthesize different medications. This will save hundreds of thousands of lives." Their effort to create a self-sustaining open-source movement for medicine has already yielded several carbon copies of mainstream pharmaceuticals and medical devices.

Part of Laufer's motivation echoes Amos Dudley's: it represents a stick-it-to-the-man effort to lower the cost of drugs and medical devices and make them more widely available. Although the attempt "to liberate life-saving pharmaceuticals from the massive corporations that own them" is applauded by some,[158] it works by evading both FDA health and safety regulations and intellectual property laws. Despite these legal risks, Laufer and his collective of fellow biohackers plow forward and continue to use information found in patent filings and academic journals to craft new DIY drugs and devices. These actions include a $30 version of the EpiPen Auto-Injector, an emergency epinephrine delivery device used to prevent deadly allergic reactions and that usually costs $600. That development prompted a warning from the FDA about Laufer's product,[159] but for the most part, the agency has not moved more aggressively against these biohackers because most of what they create is offered to the public free of charge. Again, this example reflects a strange anomaly in technology regulation that was mentioned in Chapter 1: functionally equivalent commercial and noncommercial innovations are oftentimes regulated in completely different ways. We will return to this commercial versus noncommercial regulatory problem in subsequent chapters. Regardless, we should expect more examples of evasive activities like these wherever regulations prevent distribution of valuable, life-saving goods.

Although some biohackers like Four Thieves aspire to save lives, other are just looking to enhance their own abilities for the sake of convenience or novelty. Consider the story of Jeffrey Tibbetts, an emergency room nurse who serves as a kind of chief medic for a community of biohackers and hosts the annual Grindfest in California's Tehachapi Mountains.[160] Besides sharing ideas and research, attendees often ask Tibbetts to implant small radio frequency identification (RFID) chips or magnets into their hands. He frequently obliges them, operating on them in his dedicated procedure room in his garage.[161] Or consider an Australian biohacker who got tired of digging through his pockets every time he went to board public transit. He finally removed the chip from his transport pass and implanted it in his hand, so he could simply wave his hand over the station receivers to pay for his ride. Last year, transit police fined the man for riding public transport without a pass, even though his account still had a positive balance.[162] Does he deserve this fine? Are Tibbetts and his friends criminals for choosing to slightly modify their own bodies? How are regulators supposed to find, let alone respond to, these decentralized and mostly noncommercial people or groups?

Implantable RFID chips will have many additional uses, and demand is likely to increase. Already, in Sweden, "so many Swedes are lining up to get the microchips that the country's main chipping company says it can't keep up with the number of requests."[163] Far more of those implants will likely be done professionally over time, but rogue biohacking implants probably will continue.

Some fear the dangers of biohacking, and these practices will likely remain an ongoing concern for regulators, especially as they expand to include ethically thorny issues like genetic modification and sophisticated gene-editing tools like CRISPR, which stands for Clustered Regularly Interspaced Short Palindromic Repeats, to change a segment of genetic material.[164] Concerns about unauthorized genetic modifications are growing now because of the twin threats of the pacing problem and innovation arbitrage in the form of medical tourism. "Gene-editing technology and [CRISPR], in particular, have been racing ahead even as scientists still try to sort out the ethical issues around its use," the *Wall Street Journal* noted in early 2019.[165] As discussed earlier, offshore genetic editing and fertility clinics are already on the rise, and some of them could pose risks.[166] The FDA has already moved to regulate what it regards as rogue clinics that use unapproved stem cell treatments.[167]

Perhaps the best outcome for the FDA and other policymakers hoping to more tightly police underground or offshore activities is that these efforts prove to be fringe movements that fizzle out, or that they remain deeply underground and do not come to influence health decisions by large groups of people. It is more likely, however, that biohacking and decentralized medicine will expand for a simple reason: People care deeply about improving their health and abilities. They will take advantage of new technological capabilities that let them do so—especially when those capabilities are significantly cheaper than other options. To reiterate, that does not make these technologies safe or smart, but it does mean we will need a new approach to governance as evasive entrepreneurialism expands in this arena.[168] Chapter 7 will discuss how improved risk education and awareness efforts might be one solution.

FOOD ENTREPRENEURIALISM

As noted in Chapter 1, food and plant innovation has long been riddled with social and economic controversies, mostly arising from opposition to new and better ways of growing or producing things. All too often, opposition to change comes from incumbent industries or others who have a vested interest in preserving the status quo. Alas, not much has changed over the centuries. But evasive entrepreneurs are fighting back as part of what is called a "FoodTech Revolution."[169]

For example, food truck operators have been harassed by regulators and brick-and-mortar eateries for merely selling food at the wrong locations.[170] In Chicago, city officials passed laws prohibiting food trucks from parking and serving customers within 200 feet of any traditional restaurant.[171] The effect of this onerous 200-foot rule has been to exclude food truck operators from doing any business in Chicago's famous Loop area, the prime lunchtime spot.[172] Some food truck owners decided to end service as a result.[173] Should food truck operators be prohibited from even trying to compete with traditional businesses?

Small cottage food innovators have also encountered resistance. The term "cottage food" refers to food grown or prepared by farmers and bakers on their farms or in their homes to sell primarily in their local communities. Cottage food entrepreneurs are an important part of the local food movement, and they "provide an attractive avenue to entrepreneurship for women, particularly in rural areas."[174] Unfortunately, this movement faces an "ever-growing morass of conflicting, complicated, and expensive regulatory oversight that undermines the local food movement by creating barriers between producers and consumers."[175]

But food innovators continuously find creative ways to offer the public new products.[176] Writing in 2011 about the food entrepreneur movement in Illinois, attorney Nina W. Tarr compares the situation to what happened during Prohibition:

> Historically, churches and schools might have let small food producers use their space, but the potential liability associated with allowing food production has halted such informal arrangements or driven them underground. Informal conversations with foodies around the country uncover that processed food has become the parallel to bathtub gin during Prohibition, and food entrepreneurs can be quite creative at finding ways to avoid the regulations,

such as setting up food-buying clubs that require membership to buy a home-cured ham. These arrangements have an uncanny resemblance to Speakeasies.[177]

The battle continues today. Even states that have significantly legalized food entrepreneurialism typically impose limits on advertising, location, and the overall quantity of food that can be offered.[178] Those limits can mean that holding too many bake sales or hosting a potluck is illegal.[179] Sometimes, even when a business believes it is following those rules, the interpretations change, and it becomes illegal to sell traditional salami or natural almonds.[180]

A food startup in the San Francisco Bay area was forced to shut down in 2018 after receiving pushback from local health regulators. The startup, called Josephine, was a platform to connect home cooks and those who wanted to buy a home-cooked meal.[181] Health officials stepped in to enforce laws prohibiting sales of food cooked in home kitchens, despite the startup's ability to provide great food and facilitate neighborly relationships. Does voluntarily buying a meal cooked by a neighbor necessitate cease and desist letters? Will policymakers need to pass legislation to formally allow it, as has been floated in San Francisco?[182]

Lawsuits by the Institute for Justice (IJ) and others have helped gain a bit more food freedom for those engaged in culinary entrepreneurialism. For example, a 2017 lawsuit brought by IJ successfully overturned a Wisconsin statewide ban on selling home-baked goods. A year later, home bakers and family farms across the state enjoyed new income while consumers enjoyed a range of creative new eating options.[183]

The future will get even more interesting as food production goes truly high-tech and additive manufacturing capabilities allow us to essentially replicate various foods or dietary supplements in our own homes.[184] At that point, technologically enabled disobedience will have spread all the way to the dinner table, making regulatory enforcement extraordinarily challenging.

Airspace is gradually becoming the next great platform for innovation,[185] with major tech companies and do-it-yourself entrepreneurs alike launching their drones into the skies without the prior blessing of the feds.[186] Earlier, we documented how drones were at the center of the global innovation arbitrage phenomenon. But there's even more to the story of how these devices are being used in ways that push back against onerous regulations.

Through 2016, the FAA required any commercial drone operation to be approved by the agency, yet mining companies, consumer brands, and wedding photographers, who all operated drones, never asked for the agency's permission.[187] In fact, in 2014, Rep. Sean Patrick Maloney (D-NY), a member of the House Transportation and Infrastructure Aviation Subcommittee, which oversees the FAA, hired a drone pilot to take photos at his wedding despite the FAA's near blanket ban.[188] These clear affronts to the FAA rules, and questions about the FAA's ability to enforce its commercial drone ban, forced the agency to relax its grip by issuing new drone rules that took effect in August 2016.[189] Even today, whenever people fly their own drones—even for search-and-rescue missions[190]—they are potentially in violation of dozens of FAA restrictions on unauthorized aerial activity. Should volunteer drone operators who want to help find people in the aftermath of natural disasters be fined?[191]

What about all the people who upload videos they captured with their drones? At one point the FAA was threatening people who took drone videos, although the agency eventually backed off after realizing such threats raised serious First Amendment issues.[192] Some press freedom organizations have already taken steps to defend journalists who use drones to monitor or record newsworthy events.[193]

Activists are also getting in on the act and are using drones to engage in information-gathering activities. Drones were used by professional journalists, citizen-journalists, and activists to capture footage of the 2014 protests in Ferguson, Missouri, following the shooting of 18-year-old Michael Brown by a police officer[194] and during the Standing Rock protests in North Dakota in 2016 opposing federal authorization of a pipeline in that area.[195]

In both cases, the FAA imposed Temporary Flight Restriction orders (TFRs), which temporarily handed control over airspace regulation to local law enforcement and emergency personnel, even for media-related flight activities. Freedom of Information Act requests later revealed that in both cases the FAA had granted TFRs to police after the police explicitly said they wanted to keep news helicopters and drones away.[196] Again, those actions raised clear First Amendment concerns,[197] and eventually the FAA partially reversed course and allowed a photographer to fly a drone over Standing Rock, albeit on a restricted basis.[198] The use of drones to film protests and other newsworthy activities is likely to grow and repeatedly create such enforcement headaches.

Drones are also being used in remote parts of the world to deliver medicines and vaccines for health purposes.[199] But think about more controversial efforts to use drones for personal medical reasons. Across the Atlantic, women in Northern Ireland have used drones to deliver abortion pills in areas where they are illegal.[200] This action foreshadows a day when drones might be used to courier illegal items, however defined by local authorities. There's certainly a potential risk to public safety in some of those cases, but it is equally true that other drone-related activities could advance important causes. As drone technologies continue to grow more sophisticated, we can expect controversies like these to multiply.

Had drones existed during the Revolutionary War era, leaders of the American rebellion probably would have used them to subversively drop pro-independence propaganda to the public. Instead, colonial-era rebels relied on the primary technology of freedom of that era—the printing press—even though the king of England tried to enforce restrictions on speech through licensing requirements. Ironically, "the Crown's move to stop licensing newspapers," notes American media historian Anthony R. Fellow, "led to a proliferation of them."[201] It's another example of the compliance paradox in action, and it could happen again if regulators overzealously crack down on modern technologies, including privately operated drones used for various purposes.

ALTERNATIVE TRANSPORTATION SERVICES

As documented earlier in this chapter, ridesharing services have disrupted local transportation markets in recent years and serve as a powerful example of what evasive entrepreneurialism looks like in practice. But they are hardly the only example. A transportation technology revolution is brewing.[202]

Waze, the crowd-sourced traffic application, lets drivers and commuters work together to identify traffic congestion, closed roads, police speed traps, and other things that could slow down their commutes. Although the service aims to connect local drivers and improve the quality of daily driving, some city planners and law enforcement officials dislike the fact that people use that shared information to avoid congested highways by taking side roads[203] or to identify where police cars might be hiding.[204] Waze also launched a carpool feature that allows people to hitch rides with others who may be nearby, which is essentially a variant of hitchhiking as well as competition to the ridesharing services that Uber, Lyft, and others offer.[205] The legality of all these activities is murky and depends on local laws, but it raises the question of whether crowdsourced and voluntarily shared information can or should be restricted by governments.

Meanwhile, some cities, like San Francisco and Miami, have been embroiled in a heated battle with companies that offer electric scooters for rent.[206] For a nominal fee, riders can pick up and leave the scooters just about anywhere. Scooter companies like Bird and others have followed the ridesharing model of asking policymakers for forgiveness later rather than seeking permission before launching.[207] Some local officials have not been happy with that approach.[208] But as more citizens rent or share scooters to avoid car ownership or expensive and ineffective public transportation options, should cities create permitting processes[209] and impound "illegal" scooters, as San Francisco and Miami did in the summer of 2018?[210] Such regulatory crackdowns on the micromobility revolution[211] could undermine the benefits of those scooters as an alternative to cars and buses[212] and could potentially backfire by encouraging more evasion of existing rules.[213]

Other alternative transportation options are likely to multiply, especially in urban areas, as citizens grow frustrated with inefficient government systems or restrictions that diminish the quality of life. The transportation app Citymapper uses the term "floating transport" to describe the emerging world of local transportation options available to the public.[214] Citymapper argues that transportation will be "floating" in the sense that "it has no set stops or infrastructure, and it's filling a mobility gap in our cities."[215] The options will include walking, biking, using scooters and ridesharing, and more. "While floating transport is free from many of the regulatory burdens faced by previous transport startups, it's not immune," says *Guardian* reporter

Alex Hern.[216] This tension between evasive transportation entrepreneurs and traditional transportation regulations is only going to grow in coming years.

DRIVERLESS CARS

Automobiles have occupied an important place in our society for many decades and have brought us many benefits, including a general freedom to go places and do things that would have been impossible for previous generations. Unfortunately, those benefits have had serious costs—accidents, fatalities, traffic, pollution, and so on. On average, almost 100 people lose their lives each day in America in car crashes, and 94 percent of all accidents are a result of human error behind the wheel.[217]

Autonomous vehicles or driverless cars could help alleviate many problems by diminishing the dangers associated with human error.[218] What if we were to wake up one day to find that someone had thrown a switch to give our cars greater self-driving capabilities? That is what happened on October 10, 2015, when Elon Musk, cofounder and CEO of electric car maker Tesla, announced the rollout of its Autopilot driver assistance system in its Model S vehicles. Musk announced Autopilot on Twitter that Saturday night,[219] and just a few days later, Tesla owners were able to gain access to those features via an over-the-air software update.[220]

Consider how unconventional that rollout was compared to past automotive technology debuts. In the past, a major breakthrough in automotive technology could only be obtained by purchasing a new vehicle or having the new features physically retrofitted onto your car. The latter would require taking the car to a dealership or an auto shop and then likely spending a considerable amount of money to add the new features.

With Tesla's Autopilot upgrade, by contrast, the entire process was done seamlessly at a distance via a software upgrade. That is a remarkable technical achievement, but it is also remarkable from the perspective of technological governance. Traditionally, automakers have been forced to comply with a wide variety of Federal Motor Vehicle Safety Standards that are enforced by the National Highway Traffic Safety Administration (NHTSA). Those rules govern things like seat belts, rearview mirrors, brakes, and other mechanical components in cars.[221] Technically speaking, by enabling Autopilot on its vehicles, Tesla did not violate the letter of the law as it applies to the safety regulations that NHTSA enforces, because Tesla was not making mechanical changes to its vehicles. Had Tesla removed or disabled any existing safety features on its vehicles, the company would have landed in hot water with NHTSA. But because it was simply adding a new feature, it was not in violation of any existing safety regulations.

Nonetheless, Tesla's Autopilot upgrade raised some obvious safety concerns about whether it is wise for drivers to rely on semiautonomous systems to take over the act of driving. For the purposes of our discussion, however, what is most interesting is that Musk and Tesla did that upgrade without seeking any sort of prior consent from regulators. As new features and upgrades are added to our cars instantly through software updates, what happens to the traditional regulatory system? As *The Verge* columnist Andrew J. Hawkins observes, "[T]his new reality—a reality where carmakers can introduce new features and fix system bugs remotely—raises questions about liability, safety, and the ability of historically bureaucratic organizations to keep pace with innovation in the automotive industry."[222]

Regulators have taken notice. In a nonbinding policy document published in September 2016, NHTSA claimed it "has authority to regulate the safety of software changes provided by manufacturers after a vehicle's first sale to a consumer" but also suggested that the agency "may need to develop additional regulatory tools and rules to regulate the certification and compliance verification of such post-sale software updates."[223] NHTSA expressed concern that such software changes "will substantially alter the functions and technical capabilities of those vehicles."[224]

The problem with that logic was that NHTSA was implying software updates were a problem when, in reality, they represent a great potential benefit. Because auto safety and security are an evolving baseline, we are likely to get improvements even faster with rolling software updates, with systems being refined on the fly to address real-time vulnerabilities. That sort of trial-and-error process will help autonomous vehicle companies optimize driverless car technologies for safety, security, privacy, and so on.[225] NHTSA apparently recognized that concept, because a subsequent policy guidance document from the agency dropped any discussion of regulating software updates. In subsequent guidance documents, NHTSA's parent agency, the U.S. Department of Transportation, adopted a more flexible approach that incorporated soft-law mechanisms, including voluntary guidelines, multistakeholder engagement, best practices, and other bottom-up and collaborative governance methods.[226]

Some critics say those efforts are not enough and insist that agencies continue to enforce bright-line, top-down regulations on autonomous or semiautonomous vehicles.[227] Certainly, automobile safety standards are well intentioned, and many have helped save lives. But as Chapter 4 will further document, those rules sometimes get locked in and can be slow to adapt to new realities.[228] Worse yet, they can sometimes be counterproductive and undermine sensible safety goals. The rapid evolution of autonomous vehicle technology will make it easier to eventually eliminate steering wheels, pedals, and rearview mirrors in cars, because robotic systems will not need them. What sense is there, then, in enforcing regulations mandating that all those things be in new vehicles?

Like it or not, evasive entrepreneurs will continue to put pressure on traditional vehicle regulations and force a new approach to policymaking. For example, Comma.ai provides another case study in how an emerging technology innovator used the twin threats of engaging in global innovation arbitrage and technological civil disobedience to buck regulatory threats. Comma.ai is a startup that designs a bolt-on solution to convert traditional human-operated vehicles into semiautonomous vehicles. It was founded by hacker George Hotz, who, as a teenager in 2007, gained notoriety for being the first to hack and unlock an iPhone.[229] Hotz and Comma.ai had hoped to use cheap camera and GPS technology and their own proprietary software to create a $999 after-market kit called the "Comma One."

However, in October 2016, regulators at NHTSA notified Hotz that the agency was "concerned that your product would put the safety of your customers and other road users at risk. We strongly encourage you to delay selling or deploying your product on the public roadways unless and until you can ensure it is safe."[230] Hotz escalated the controversy by reposting the full letter online and responding angrily to it via Twitter, decrying the agency's "threats" and the absence of an "attempt at a dialog (sic)."[231] In two additional tweets that followed, Hotz said he would "rather spend [his] life building amazing tech than dealing with regulators and lawyers"[232] and would be canceling the Comma One in the United States and that his firm would "be exploring other products and markets. Hello from Shenzhen, China."[233]

Hotz's threat to leave the United States and embrace a global innovation arbitrage response drew a great deal of media coverage,[234] but the firm quickly abandoned the plan to leave and instead announced that it would be open-sourcing its software and offering it freely to other developers.[235] In this way, Hotz was engaging in a rather creative form of technological civil disobedience: making it harder for regulators to control the technology by removing himself and his firm as gatekeepers of it.

The examples of evasive entrepreneurialism set by Elon Musk and George Hotz raise the question of just how far safety regulations can be stretched before a major backlash ensues. Following a handful of crashes that involved driverless cars in 2017 and 2018, including one in Arizona that resulted in a fatality, some critics pushed back against driverless cars and insisted that they should not be allowed on public roads until they had gone through traditional regulatory processes. But that process has stalled for driving safety improvements. Basically, today's auto safety laws are trying to squeeze a little more juice out of an old lemon when what we really need is an entirely new fruit salad of options. Only bold innovations will get us the sort of moonshot innovations we really need.

Meanwhile, America's patchwork of vehicle regulations has scared off some developers of autonomous vehicle technology such as Audi, which in 2018 offered European drivers far more robust driverless tech than Americans were given access to.[236] This action illustrates how global innovation arbitrage will likely continue to be a major factor in driverless car policy battles to come.

BLOCKCHAIN AND DECENTRALIZED MARKETPLACES

Financial transactions have traditionally been routed through some central authority, such as a bank, payment processor, or credit card company. All of those entities are highly regulated. The rise of the first decentralized cryptocurrency, Bitcoin, and the distributed blockchain technology that it introduced allowed individuals to transfer value directly to recipients without needing to rely on a trusted third party in the middle. Because there is often no central entity that can be regulated, Bitcoin and blockchain technology pose unique challenges for a wide variety of tax and regulatory policies and systems.[237]

Bitcoin began with an act of technological civil disobedience. In announcing its launch on February 11, 2009, the unidentified programmer known as Satoshi Nakamoto declared that Bitcoin would address some of the core problems associated with traditional government-backed or regulated currencies:

> The root problem with conventional currency is all the trust that's required to make it work. The central bank must be trusted not to debase the currency, but the history of fiat currencies is full of breaches of that trust. Banks must be trusted to hold our money and transfer it electronically, but they lend it out in waves of credit bubbles with barely a fraction in reserve. We have to trust them with our privacy, trust them not to let identity thieves drain our accounts. Their massive overhead costs make micropayments impossible.[238]

Bitcoin would offer an alternative in the form of "a global distributed database, with additions to the database by consent of the majority, based on a set of rules they follow," Nakamoto said.[239] This publicly distributed ledger technology is called the blockchain. Blockchain allows users from across the globe to interact through a decentralized peer-to-peer network and create not only alternative currency systems, but also other services that blockchain ledgers can be used to verify.[240]

Ten years after Nakamoto launched Bitcoin, we still have no idea who he (or she) is, or if it is even a single person behind the alias. But we know the result of this revolutionary idea. Bitcoin market valuation soared to more than $100 billion, and multiple alternative cryptocurrencies have launched to compete against it.[241] Regulators have been scrambling to keep pace ever since.

The IRS ruled that cryptocurrencies were to be taxed like property in 2014,[242] but not before hundreds of thousands of Bitcoin transactions were conducted.

Even after IRS efforts to crack down on cryptocurrency users, many of them flaunt U.S. tax law.[243] Although the current base of cryptocurrency users is small relative to the general population, if more people flock to cryptocurrencies, it could prompt governments to reassess the burden of their current tax policies.[244]

The logic of distributed computing and distributed networks is being extended even further. Even Facebook entered the fray in mid-2019 with the launch of Libra, a distributed ledger-based digital currency system that, though more permissioned than Bitcoin, generated immediate opposition from many lawmakers and executive branch officials who claimed it posed national security concerns.[245]

Meanwhile, similar to the way that Bitcoin introduced a distributed currency to the world, other developers have begun working on creating distributed marketplaces for goods, services, and information. For example, the OpenBazaar distributed marketplace project employs cryptocurrencies and general distributed computing to create a platform that cannot be controlled or shut down by any central operator.[246] Rather than requiring a central authority, like eBay or Amazon, to run the platform and prune buyer and seller accounts, OpenBazaar aims to place that power solely in the hands of users. Trustworthiness and quality will be surmised through user-given ratings and feedback. The buyer (and seller) must beware: the OpenBazaar administrators would not be able to reverse a transaction or sanction any bad actor. But on the flip side, neither governments nor corporations have the power to control or shut down the peaceful exchange of any good or service, including ones they would like to clamp down on.

Similarly, distributed computing can facilitate markets that primarily trade in *information*, such as prediction markets or gambling operations.[247] Although the Commodity Futures Trading Commission heavily regulates such undertakings,[248] spirited developers have nonetheless plowed forward with projects to open prediction markets, which allow self-styled soothsayers to bet on the likelihood of upcoming events occurring in some way or at all. The market price of such individual options is a kind of incentive-compatible probability that reality will unfold in one way or another.[249] One example of such a project is Augur, which is currently operating in beta on the Ethereum network.[250] Ethereum is a smart contracting platform that employs blockchain applications to allow users to program contracts and legal arrangements through code. The structure of such arrangements could catalyze a new generation of business relationships and programmed incentive structures, yet many of those projects may run afoul of established Securities and Exchange Commission regulations and eventually be prosecuted as illegal securities.[251]

Less socially useful, perhaps, are distributed gambling outposts,[252] which can similarly be arranged in such a way that they cannot be shut down easily, if at all. Policymakers looking to crack down on existing pseudo-information markets and gambling platforms—such as popular online fantasy sports facilitators[253]—would be wise to consider the extent to which harsh policies may merely quicken the pace of impossible-to-shut-down alternatives. For example, in 2014, NBA commissioner Adam Silver wrote that sports gambling, "should be brought out of the underground and into the sunlight where it can be appropriately monitored and regulated,"[254] and a 2018 U.S. Supreme Court decision has opened the door for states to legalize sports gambling.[255]

Those examples point to the continuing revolutionary potential of blockchain technologies and distributed applications to upset traditional regulatory arrangements and technological governance norms. As my colleague Tyler Cowen observes, "a blockchain is actually a form of governance and that is what makes it such a potentially radical idea."[256] How far could the implications reach? "Will the next generation of tech be done by legal systems and corporations or by blockchains? Or some combination of those? That's the new intellectual battlefield," Cowen says.[257] Indeed, it is a *policy* battle as well. The wrong kind of policy posture may very well force some of those projects to shut down prematurely, but it may prompt other, more ambitious (or perhaps more unscrupulous) outlets to arrange their code in such a way that it cannot be traced or shut down at all. For example, the "darknets" of recent years allowed semianonymous transactions to take place at a distance, including in some illegal products. But darknet transactions were often technically complicated and risky (especially because they relied on traditional postal service or parcel delivery mechanisms, which made products more traceable).[258] Darknet markets were also susceptible to theft by market operators or capture by law enforcement. Those facts have led to the evolution of still other gray or black markets that are even more decentralized, such as "dropgangs," which incorporate dead drops (hidden deliveries made in public places).[259] Dropgangs forgo the use of a central platform at all, opting instead to connect through encrypted messaging platforms.[260] Sellers and buyers negotiate transactions with encrypted messaging services and then use Bluetooth beacons, GPS coordinates, and code words to surreptitiously complete deals.

It remains unclear whether such highly decentralized markets will ever gain enough widespread consumer trust to scale up and become prevalent enough for policymakers to attempt more serious crackdown efforts. It seems unlikely, especially if they are seen only as outlets for risky or illegal activities. Fraudulent activities deserve greater attention and policing. Nonetheless, it is clear that such new technological capabilities and evolving marketplace mechanisms will make various types of evasive entrepreneurialism even more likely in coming years.

Spontaneous Deregulation?

The case studies discussed here illustrate how evasive entrepreneurialism and technological civil disobedience are on the rise in many different sectors. What are the ramifications for the future governance of those sectors?

One result might be the spontaneous deregulation of certain technologies or sectors. Typically, the deregulation of a sector requires affirmative steps be taken by legislative bodies or regulators to formally remove restrictions on innovation or new entry. By engaging in acts of technological civil disobedience, however, evasive entrepreneurs can sometimes break free of traditional regulatory restraints or avoid the imposition of new ones. If policymakers do not respond or, as discussed in the next chapter, if they engage in *rule departure*—a phenomenon where, for whatever reason, they refuse to enforce existing laws and regulation— then that sector could become effectively deregulated.

This de facto rather than the de jure elimination of traditional laws and regulations can be thought of as spontaneous deregulation.[261] "Benign or otherwise, spontaneous deregulation is happening increasingly rapidly and in ever more industries," argue Benjamin Edelman and Damien Geradin in a *Harvard Business Review* article on the phenomenon.[262] This situation means that many laws and regulations may remain on the books long after anyone—including enforcers—cares about them. The next chapter begins by discussing a number of laws that remain on the books many decades after their implementation, yet the public ignores them and policymakers choose not to enforce or update them.

Spontaneous deregulation has interesting implications for governments, innovators, and average citizens. Governments will have to consider whether and how to clean up their regulatory messes if innovators and the public choose to ignore certain regulatory policies. For entrepreneurs, the trend means that, strategically speaking, they will increasingly have the ability to consider whether to formally petition policymakers for regulatory liberalization or to instead roll the dice as regulatory entrepreneurs in pursuit of spontaneous deregulation. And for average citizens, to the extent that they want to try their hand at homegrown sorts of free innovation—including noncommercial innovation—they may now be emboldened to undertake creative endeavors that might have been too legally risky in earlier decades.

However, even as evasive activities challenge regulatory regimes, some new technological capabilities that could enable fraudulent or harmful behavior will still need to be policed. Finding sensible governance strategies will be a major focus of the second half of this book.

Evasion in Born Free versus Born Captive Sectors

Evasive techniques and spontaneous deregulation will not always play out smoothly or be effective, however, particularly regarding some heavily regulated sectors. The prospects for successful evasion are inversely related to the degree to which a sector or technology is born into what we might think of as regulatory captivity. It is useful to contrast born-in-captivity technologies with those that are born free to understand why successful examples of evasive entrepreneurialism will be much less likely in some sectors.

Consider computers and most modern digital information technologies. At least here in the United States, those technologies were mostly born free.[263] That is, innovators in those fields did not face preexisting laws and regulatory regimes that might have limited their ability to experiment with new ideas. Beginning in the mid-1990s, permissionless innovation became both the marketplace and the public policy norm for the digital economy.

By contrast, earlier information technologies from the analog era (i.e., broadcasting, cable, and telephony) were born captive. Those older media and communications providers were confronted with a veritable alphabet soup of regulatory agencies. Technocratic regulators in those agencies viewed each new product or service with suspicion and required anyone with a new idea to first fall in line with the Mother-May-I regulatory regime that treated innovations as essentially guilty until proven innocent.

With the rise of the internet and digital technologies, however, the public came to enjoy innovations that were generally born free of such preemptive constraints. For example, websites and blogs, social networks, smartphone apps, and other digital technologies have mostly stayed in the born-free camp. Other technologies have been lucky enough—either through intention or plain luck—to develop in an environment of policy restraint. For example, robotics, 3D printing, and virtual reality have, at least thus far, been born free because there is no Federal Robotics Agency, 3D Printing Act, or Virtual Reality Commission.

Other technologies and sectors have not been so lucky and have been born captive in that they were quickly subsumed by laws and regulatory bodies or pigeonholed into old regulatory regimes. Financial technology, or fintech, is probably the best example. New financial technologies are confronted with a minefield of overlapping laws and agencies at the federal, state, and global level. "Every example of fintech is highly regulated from the moment it is conceived of," says Brian Knight, a Mercatus Center expert on financial regulation. "In some cases, the existing regulatory environment harms innovation by forcing firms to comply with multiple, often inconsistent sets of rules, pay the costs of having to constantly monitor numerous state and federal regulators, and face the uncertainty of not knowing whether an activity is subject to regulation."[264] The best hope for fintech innovators is probably jurisdictional competition that allows them to choose more hospitable regulatory arrangements.[265] But innovation arbitrage in this sector tends to be more difficult than in other fields.

Aviation and space policies are also encumbered by a stifling array of overlapping federal agency prohibitions before anything can, quite literally, get off the ground.[266] Likewise, experimentation in the electricity and energy marketplace is stifled by a complex assortment of federal and state regulations that make innovation exceedingly difficult.[267] The prospects for evasive entrepreneurialism and spontaneous deregulation in such sectors are dimmer than they are for AI, 3D printing, virtual reality, and others that are generally born free of preexisting regulatory regimes.

But context matters. In the field of airspace regulation, for example, the chances are quite slim that supersonic transportation innovators can act in a more evasive manner, compared with small-drone innovators who might be able to do so. By its size and nature, a supersonic jet makes an easy target for regulation. After all, those jets are large, need more space for taking off and landing, and probably will need to be near major metropolitan areas to attract passengers and be profitable. By contrast, drones are smaller, unmanned, and can be operated almost anywhere. The chances for drone operators to evade regulation are greater, therefore, and we have already discussed several examples of technological disobedience in which they are involved.

When we consider the likely success of evasive or disobedient strategies in any given context, we should keep such nuances and realities in mind.

3

DISOBEDIENCE: THEN AND NOW

s it really accurate to refer to acts of evasive entrepreneurialism as civil disobedience? I will argue that it is accurate so long as one adopts a broader understanding of civil disobedience and appreciates how many forms of evasive entrepreneurialism flow out of a long-standing American tradition of dissent more generally. We will also consider the morality of evasive acts and whether they can be justified.

This chapter will also consider what motivates rule departures, which occur when government officials choose not to enforce public policies for various reasons, often because they object to the morality or sensibility of those policies. We can expect to see more of that behavior by public officials in coming years, for reasons I will identify here and in the following chapter.

Why Do Some Choose to Disobey the Law?

Whether or not there is an affirmative duty to obey the law is "not only among the oldest and most enduring of philosophical questions, but a question that has continued to attract the attention of legal, social, and political philosophers."[1] Why is it that people sometimes choose to disobey the law?

It should be noted that the vast majority of citizens obey most laws—at least to the extent that they understand what the law says and means. In his 1990 book, *Why People Obey the Law*, Tom R. Tyler wrote that, generally speaking, citizens obey most laws not out of a fear of sanction, but more out of a sense that the laws are legitimate—either by their nature or at least in the way they were formulated.[2]

But sometimes people push back against the law—in both small and large ways—for many reasons. Refusal to obey law is sometimes rooted in matters of personal morality. Freedom of conscience and moral repugnance regarding existing state policies often lie at the heart of concerted opposition to existing or proposed laws and regulations. Proponents of civil disobedience usually speak of their activities as a moral duty, as the minister Jonathan Mayhew did in his sermons and writings leading up to the American Revolution. In his 1750 "Discourse Concerning Unlimited Submission and Non-Resistance to the Higher Powers," Mayhew argued that "a regard to the public welfare, ought to make us withhold from our rulers, that obedience and subjection which it would, otherwise, be our duty to render to them."[3] Mayhew continued:

> If it be our duty, for example, to obey our king, merely for this reason, that he rules for the public welfare . . . it follows, by a parity of reason, that when he turns tyrant, and makes

his subjects his prey to devour and to destroy, instead of his charge to defend and cherish, we are bound to throw off our allegiance to him, and to resist. . . ."[4]

Using similar reasoning, Benjamin Franklin proposed that the Great Seal of the United States include the phrase "Rebellion to tyrants is obedience to God." His proposal was rejected, but that sentiment reflected the ethos of the revolutionary era, and it inspired Thomas Jefferson and others who signed the Declaration of Independence.[5] That sentiment is hardly surprising considering the many acts of civil disobedience leading up to the Revolutionary War, including extensive smuggling and pamphleteering in opposition to repressive governmental policies.[6]

Roughly a century later, in his influential 1849 pamphlet that helped popularize the term "civil disobedience," Henry David Thoreau argued, "It is not desirable to cultivate a respect for the law, so much as for the right. The only obligation which I have a right to assume is to do at any time what I think right."[7] Another century later, in the famous "Letter from a Birmingham Jail," the Rev. Martin Luther King Jr. similarly insisted, "One has a moral responsibility to disobey unjust laws. I would agree with St. Augustine that 'an unjust law is no law at all.'"[8]

These writings make it clear that civil disobedience has deep roots in the American political tradition. Civil disobedience has been a prominent feature of social and political movements in the United States to abolish slavery, advance various civil rights, oppose war or military conscription, promote freedom of speech and expression, and so on. Again, those engaging in disobedience to advance such causes have stressed that disobeying the law is morally justifiable because public policies were unjust and deeply offensive to values they held dear. The disobedient also typically resist those policies in a public way and use forms of nonviolent protest to defy rules or regimes they find repugnant. The disobedient are also usually willing to pay some sort of price (imprisonment, fines, etc.) for their actions.

Different Flavors of Disobedience

Advocates of civil disobedience and scholars who have written about it sometimes disagree about the specific elements or attributes of civil disobedience, however. As one scholar in the field has noted, civil disobedience "is not a self-explanatory term" and "there are several 'grey' areas to think about."[9] Advocates of civil disobedience and their particular causes also come in many different flavors.

Left-leaning advocates of civil disobedience, like historian Howard Zinn, focus on how the public can act in concert to encourage the state to act forcefully in advancing various "social justice" objectives.[10] Although this effort can entail ending or curtailing some state powers—such as voting restrictions or military conscription—it can also involve expanding government powers and programs in the name of redistributing wealth or launching new regulatory initiatives. For example, the 2011 Occupy Wall Street movement, which featured protests in New York City and other cities, was focused on "anger at the global financial system, corporate greed and government cutbacks."[11] The goal of civil disobedience in this instance was the expansion of state control over the economy, or at least the financial sector. More recently, some left-leaning scholars have called for more targeted types of technologically enabled disobedience in service of expanded individual privacy. In their 2015 book, *Obfuscation: A User's Guide for Privacy and Protest*, Finn Brunton and Helen Nissenbaum advocate a "limited revolution" through "the deliberate addition of ambiguous, confusing, or misleading information to interfere with surveillance and data collection."[12] Their goal is to use obfuscation technologies to counter not only government data collection and surveillance efforts but also those of private entities.

By contrast, more libertarian-minded defenses of civil disobedience can be found as far back as the mid-1500s in Étienne de La Boétie's *Discourse of Voluntary Servitude*,[13] and then again recently in Charles Murray's 2016 book, *By the People: Rebuilding Liberty without Permission*. Murray proposed "a declaration of limited resistance to the existing government" because, he said, it "has in many respects become destructive of our unalienable rights," and "has lost elements of its legitimacy."[14] The goal of civil disobedience for many libertarian advocates is a general diminishment of state power over society and the economy alike. With the rise of personal computing and the internet, some libertarian-minded tracts, such as Timothy May's 1992 "Crypto Anarchist Manifesto"[15] and John Perry Barlow's "A Declaration of the Independence of Cyberspace,"[16] suggested that new digital technologies could help challenge the authority of governments in various ways.

My goal here is not to develop a strict taxonomy or hierarchy of civil disobedience that identifies which forms of it are more important than others. But I do believe that, regardless of the specific ends sought by those engaging in disobedience, *there is inherent value in the actual process of dissenting more generally.* Dissent and disobedience have been an important part of the American tradition of freedom of speech, expression, and voluntary association. They should also be seen as a healthy part of keeping a representative democracy working properly or, more simply, just making sure that public policies reflect common sense and align better with the evolution of social, cultural, and economic values. More specifically for the purposes of this book, dissent and disobedience are perhaps best viewed as useful correctives when laws and the regulatory system become untethered from new realities and begin to lose their sense of legitimacy with the public.

How, if at all, are traditional conceptions of civil disobedience and more modern technologically enabled forms of disobedience related? The philosopher John Rawls identified the core elements of civil disobedience in his much-cited definition of the term as "a public, nonviolent, conscientious yet political act contrary to law usually done with the aim of bringing about a change in the law or policies of the government."[17] Many of the examples of technological civil disobedience we are witnessing today are similar to traditional forms of civil disobedience in that they tend to be *public* and *nonviolent* in character. However, some of the examples of evasive entrepreneurialism discussed here are not always as *conscientious* or *deliberately unlawful* as past forms of disobedience. Sometimes people just use technology to accomplish things without even realizing the gravity of their act or that their actions might be illegal in some fashion. In essence, they are informally opting out of legal regimes that they do not fully understand. Although it may be more accurate to refer to this type of action as legal avoidance or legal indifference, I will continue to use the term "technological civil disobedience" to describe most of these activities, but I acknowledge that important differences exist between it and traditional forms of disobedience.

Here and in Chapter 5, I will extend this frame of thinking and argue that innovative acts can be viewed as mini-rebellions or marginal revolts that can help bring about positive change through small acts of technologically enabled creativity and resistance. By giving citizens additional voice and exit options beyond politics to protest or push back against policies they disagree with, innovative activities help citizens live lives of their own choosing. Further, I will argue that these activities have a potentially beneficial effect on the administration of various government policies. In essence, innovation can act as checks and balances that encourage a realignment of laws and regulations that have failed to keep pace with the times or have grown burdensome, unwieldy, inefficient, or even unjust.

Liberty of Association and the Spirit of Resistance

Writing in the early 1970s, the political theorist Hannah Arendt argued that, properly understood, civil disobedience represented "nothing but the latest form of voluntary association, and . . . thus quite in tune with the oldest traditions of the country."[18] Arendt built her approach around Alexis de Tocqueville's observation in the 1830s in *Democracy in America* that "the liberty of association has become a necessary guarantee against the tyranny of the majority."[19] Arendt suggested that when the public freely engages in such associations and acts in concert to advance various ends, *especially acts of dissent and disobedience*, it has a salubrious effect on democracy.

This thinking was not radical, although it might sound that way to some today. Recall Thomas Jefferson's famous assertion that "a little rebellion now and then is a good thing, and as necessary in the political world as storms in the physical."[20] He argued, "God forbid we should ever be 20 years without such a rebellion," because "what country can preserve its liberties if their rulers are not warned from time to time that their people preserve the spirit of resistance?"[21]

Jefferson's call for periodic resets of government would entail repeated revolutionary acts that would be hard to accomplish and highly disruptive to society and the economy alike. Civil disobedience of the sort Arendt envisions would stop short of such sweeping actions but could still help keep government actors on their toes and responsive to the people. Arendt claimed that "when the established institutions of a country fail to function properly and its authority loses its power" it creates a sort of "emergency" for the health of the republic.[22] "Ever since the Mayflower Compact was drafted and signed under a different kind of emergency," she argued, "voluntary associations have been the specifically American remedy for the failure of institutions, the unreliability of men, and the uncertain nature of the future."[23]

As I already noted and will discuss further below, much modern law simply no longer reflects the will of the people or fails to align well with social values or technological realities. "When the law differs from the socially desired level of regulation, we might observe a backlash in which society tries to correct the law and align it with current social norms," observe Emanuela Carbonara, Francesco Parisi, and Georg von Wangenheim.[24] They describe this dynamic process and how a misalignment of law and social values leads to civil disobedience and then recalibration of public policy:

> When the law exceeds what is deemed socially acceptable, people protest, adopt a behavior of civil disobedience, and try to stop its enforcement. In general, they approve an observed infringement of the "unjust law," and they try to stop the application of too-severe penalties.[25]

If public officials double down on enforcing the offending policies, it does not mean the public will submissively fall in line. Instead, dissent and disobedience might simply increase in response. "These unintended effects of legal intervention would thus lead to a compliance paradox—an increase in the strictness of the law would lead to an increase in legal violations, defeating the goals pursued by the lawmaker," they say.[26]

How Dissent Advances Government Accountability

That concept is a useful frame for understanding why technological civil disobedience is on the rise today—and also why it is worth defending.

Not every act of civil disobedience (including many of the examples of technological civil disobedience discussed throughout the book) is necessarily undertaken to advance a profound cause. Sometimes citizens engage in acts of disobedience for less weighty reasons—and in ways that are more indirect—because they are frustrated with the status quo or feel that existing governance arrangements have broken down. Arendt asserted that civil disobedience sometimes arises because "a significant number of citizens have become convinced . . . that the normal channels of change no longer function, and grievances will not be heard or acted upon."[27]

In the same way, many citizens today feel that they are not getting meaningful representation from our supposedly representative government. Public trust in government has been in decline for the past four decades and has reached historic lows. According to the Pew Research Center, "Only 18% of Americans today say they can trust the government in Washington to do what is right 'just about always' (3%) or 'most of the time' (15%)."[28] Such results suggest that government today has lost legitimacy with an overwhelming number of citizens.

Of course, the quest to calibrate or obtain the consent of the governed is no easy task in a pluralistic society.[29] As subsequent chapters will document, however, modern governance systems have grown increasingly incomprehensible, inefficient, and illogical. As those systems have done so, many citizens have grown dissatisfied with politics on so many levels that they are finding novel ways to push back against the lumbering leviathan that is our modern government. Moreover, they now possess tools or capabilities that make it increasingly easy for them to ignore or evade certain laws or regulations of that sort.

The defense of the various forms of technological civil disobedience presented in this book rests on this foundation and is rooted in the simple belief that *sometimes shaking things up a bit is not such a bad thing.* "To the extent that innovation or legal change can benefit society, some of this activity has the potential to provide social value," argues Elizabeth Pollman of Loyola Law School.[30] Thus, "when entrepreneurial activity produces useful technology or innovative services that improve citizens' well-being, productivity, or quality of life, a justification may exist for the disobedience that goes beyond a simple cost-benefit analysis of a corporation's gain," she says.

This idea of shaking things up can help improve government processes. Dissent in all its forms has some inherent value that is worth protecting in a representative democracy that purports to care about pluralistic values and the will of the people. When innovators circumvent captured bureaucracies and disrupt inefficient regulatory regimes that have failed to serve the public interest, the changes can enhance social welfare by expanding choices and encouraging positive reforms.[31] Pollman notes that "there is informational value in disobedience that points to areas of law that might be outdated, that conflict with the law of other jurisdictions, or that reflect a political or market failure."[32]

In other words, if technologically enabled acts of dissent and disobedience can help citizens push back against *un*representative government and broken policies that impinge upon individual rights, then those activities are valuable and worth defending. This process can help us recalibrate government policies and institutions to ensure that they are once again viewed as legitimate and in line with the public interest and modern social, economic, and technical realities.

Should Profit Seeking Matter?

But should entrepreneurial activities of an evasive nature be equated with civil disobedience? Some might object to the inclusion of profit-seeking activities in any definition of civil disobedience. It would be bizarre, however, to suggest that the simple act of being entrepreneurial or seeking to earn a living somehow means that such acts cannot also represent forms of civil disobedience.

Scholars have long recognized that entrepreneurs can act as agents of social and cultural change without engaging in monetary transactions.[33] Entrepreneurial activities can also quickly transition from one (noncommercial) to the other (commercial). Many of the examples of technological civil disobedience cited throughout this book have both noncommercial and commercial variants; others begin as noncommercial activities but then become commercial ones. The opposite can also occur, as we saw with the George Hotz example in the previous chapter. He originally aimed to sell his autonomous driving technology but then decided to open-source the technology, giving that information to the world freely.

Noncommercial versions of some activities are often left largely, if not completely, unregulated. But as soon as equivalent commercial opportunities arise, either an existing regulation kicks in or policymakers seek to apply new regulation to the activity in question. Does that make any sense? Also, what will happen as new technological capabilities make it easier to create and share noncommercial devices and services?

Differential treatment of functionally identical activities is often premised on the belief that the commodification of an activity somehow sullies a previously noble, or at least innocuous, undertaking or that the commercial variant of an activity is somehow more risky than the noncommercial. Both rationales for regulation are misguided.

The first rationale is highly subjective because whether one regards a commodified activity as less respectable or dignified than noncommodified equivalents is mostly a matter of personal morality and taste. Generally speaking, regulation should not be premised upon such subjective considerations.

Recall the examples cited earlier about how some parents are using 3D printers and open-source software designs to create DIY diabetes management tools for their children by coding their own insulin management software, as well as the volunteers who use the same tools to make prosthetic hands for children with limb deficiencies. If people voluntarily use such technologies to make devices that are potentially in violation of FDA medical-device regulations, should it make any difference whether they did so for free or asked for a small fee?

Apparently, it would make a difference, because according to a statement posted on its website in November 2017 titled "Information about Self-Administration of Gene Therapy," the FDA said,

> FDA is aware that gene therapy products intended for self-administration and 'do it yourself' kits to produce gene therapies for self-administration are being made available to the public. The sale of these products is against the law. FDA is concerned about safety risks involved. Consumers

are cautioned to make sure that any gene therapy they are considering has either been approved by FDA or is being studied under appropriate regulatory oversight.[34]

Notice what the statement does and does not say. The FDA clearly establishes that the commercial sale of any such products is strictly regulated. On the other hand, the statement is ambiguous regarding whether or not noncommercial DIY gene therapies would be subjected to the same level of scrutiny or sanctions. Does that make any sense? And what happens when the lines get blurred between commercial and noncommercial innovative activities?

For example, what if the creators of the 3D-printed prosthetics or DIY gene therapies ask for donations, or receive support from a charitable foundation to cover their operating costs? Or, what if they begin undertaking those activities for free, but then later decide to form a corporation and sell the prosthetics or gene therapies for profit? Presumably that would trigger regulatory sanctions. In these hybrid hypotheticals, innovators are potentially in violation of the law, whether they are doing so to protest those policies intentionally or not. How much should it matter how they recoup the costs associated with providing those services? If an act is technologically disobedient (or illegal) when done for free, then when the identical act is done for a fee, it still represents an act of technologically enabled civil disobedience and probably should also be illegal.[35]

Critics might also suggest that evasive entrepreneurs do not have as much to lose as those individuals and groups who have engaged in acts of civil disobedience in other contexts, especially noncommercial ones. To the contrary, many evasive entrepreneurs potentially have *even more* to lose. Like others who engage in acts of disobedience, evasive entrepreneurs put their liberty and property at risk. They might get fines or jail time for their actions, but they also risk hurting their personal or organizational reputation with the public, putting their employees or shareholders at risk, or losing their organization or business entirely. That is clearly the case with many of the examples cited in the previous chapter.

It is also not necessarily true that the commodification of a previously noncommercial activity makes it riskier in some fashion, hence requiring greater regulatory scrutiny. In fact, that is quite likely false in most cases. First, commercial actors have strong reputational incentives to be good market actors. Firms face much greater public and press scrutiny than noncommercial actors. Commercial operators seek to maintain brand loyalty to earn new or repeat business. Second, commercial actors also possess strong incentives to avoid stiff legal liability that noncommercial operators might not face. Firms are more likely to be quickly dragged into court at the first sign anything might be wrong, even when there is no merit to the claim. Third, commercial operators are more likely to carry insurance to cover risks associated with their activities.

Morality Questions—Ex Ante versus Ex Post

Finally, critics might claim that the people engaged in acts of civil disobedience traditionally have had much more important objectives, whereas evasive entrepreneurs are not engaged in such noble pursuits. Yet trying to earn a living and being inventive are extraordinarily important pursuits that deserve more moral weight than they are sometimes given.[36] Offering the public new and better goods and services is a fundamental driver of human betterment, and innovative acts should be appreciated for the way those acts expand the range of choices that the public has at its disposal.[37] Moreover, evasiveness or technological disobedience—whether commercial or noncommercial—can help make government more accountable to the people.

To be sure, some forms of technological disobedience—especially by innovators seeking to eventually make a profit—will remain controversial. Heated debates will continue to take place about where to draw the line between ethical versus unethical forms of disobedience.[38] "The welfare effects of specific cases of evasive entrepreneurship can be more or less easy to evaluate, but the basic philosophy for doing so is easily understood," note Niklas Elert and Magnus Henrekson. "However, welfare analysis is not the only standard for judging the effects of evasive entrepreneurship. Other moral and ethical considerations must also be reckoned with when evasive actions are judged."[39]

Everyone would agree, for example, that evasive activities built on fraudulent claims or practices would harm consumer welfare. When blood-testing startup Theranos broke on to the scene in the mid-2000s, investors showered the firm and its founder Elizabeth Holmes with hundreds of millions in support, and great media fanfare followed for Holmes and the firm throughout the 2010s. Holmes promised to revolutionize blood testing by making it cheaper and more convenient for consumers, but she and Theranos were eventually exposed as fraudsters making baseless assertions about nonexistent technology. The firm came under investigation by a variety of regulatory authorities, was subjected to lawsuits, and then was driven into bankruptcy and shut down entirely in 2018. Although Theranos was not the same sort of evasive entrepreneur described throughout Chapter 2, the firm still serves as an example of why laws will always play an essential role in policing activities that could bring harm to the public.

Even if entrepreneurs think they are serving the public good, they may not be. Should innovators choose to engage in evasive techniques to push back against laws and regulations they think are misguided or unjust, there could be significant penalties associated with their actions, including fines, lawsuits, or even jail time. In many cases, however, their gambits will work and the public will be provided with a new and better good or service that enriches consumer welfare.

This is the irony associated with evasive entrepreneurialism. It is easy to justify evasive acts after they successfully usher in an innovation that most people come to regard as life-enriching. In fact, in many such cases, people end up wondering why we didn't have access to such things long before. But it is much more difficult to justify the evasive acts that led to such innovations preemptively. Before or as evasive acts take place, many people will question the morality of innovators pushing back against the law or making an end-run around the law. But those same people will likely later defend evasiveness that seemingly served the public good. Thus, the welfare calculus associated with evasive entrepreneurialism will often become a consequentialist exercise that few defend ex ante but many defend ex post.

We will return to these ethical and legal considerations in Chapters 6 and 7, but the morality of evasive entrepreneurialism remains ripe for more robust study, and it will no doubt receive more attention as such activities multiply.

Rule Departure: Government Civil Disobedience?

It is worth noting that public officials sometimes engage in acts of legal evasion or disobedience of their own, although they are almost never labeled as such. The term "rule departure" refers to "deliberate failures, often for conscientious reasons, to discharge the duties of one's office."[40] Civil disobedience by public officials can be defended on many of the same grounds as private disobedience can. A judge or jury that refuses to convict someone despite that person clearly being in violation of the law is an example of rule departure. For example, judges have refused to punish draft dodgers, and juries have sometimes nullified laws that they find unconscionable.

An entire legislative body can also engage in rule departure. When state lawmakers refuse to comply with federal marijuana restrictions because officials in those states favor decriminalization, that refusal represents rule departure between levels of government.[41] Similarly, in May 2018, Vermont became the first state to legalize importing prescription drugs from Canada in an attempt to gain access to lower-priced drugs for its citizens.[42] That move departed from federal rules, which regulate the importation of drugs into the United States. City governments also engage in rule departures.

Rule departure is often even more localized. When the Trump administration took office and announced more stringent immigration rules, many mayors and local officials across America announced that they would become sanctuary cities and not follow federal immigration reporting requirements.[43] More recently, some county officials, including local sheriffs, are part of a Second Amendment sanctuary movement that resists state or federal gun restrictions.[44] In the former case, the local officials are mostly liberal; in the latter, they are mostly conservative. The only thing unifying those officials is a commitment to engage in rule departure for moral reasons.

For purposes of our discussion throughout this book, rule departure by regulators is worth monitoring because it seems to be increasing in many fast-moving technology sectors. Agency regulators engage in rule departure for a variety of reasons. When a regulator turns a blind eye to new developments that might run afoul of existing rules, it is sometimes referred as *agency discretion* or *forbearance*. Sometimes such discretion or forbearance is permitted by the agency's authorizing statute or subsequent laws. Other times, agency officials are simply bending or ignoring the law in creative ways to make things work as best they can as the pacing problem rears its head with increasing regularity.

Consider the FAA's approach to commercial drones as they became more prevalent. Unless amateur drone pilots do something egregiously stupid—such as flying a drone over a major league baseball park—the FAA seems to have largely given up on enforcing the agency's drone licensing requirements, which have been in effect for commercial operators since 2016.[45] In the first 18 months the licensing requirements were on the books, only one drone pilot was issued a warning.[46] In response to inquiries, the agency typically says it "strongly encourages compliance" with its permitting requirements, but because, as Gizmodo's Melanie Ehrenkranz notes, "passing FAA's test to receive a drone pilot's license is reportedly a giant pain in the ass," and is also costly, most drone owners don't bother complying.[47]

Why isn't the FAA trying harder to enforce its own rules? After all, *it's the law*. It might be because the agency recognizes that stronger enforcement could bring about the compliance paradox discussed earlier in which "an increase in the strictness of the law would lead to an increase in legal violations, defeating the goals pursued by the lawmaker."[48] Or perhaps the cost of going after more than just a few problematic users would simply be too high for the agency, which must decide the best way to spend limited resources. Limited resources likely are an even more binding constraint on state and local authorities, which the FAA has said "are often in the best position . . . to stop unauthorized or unsafe UAS [unmanned aerial system] operations."[49] Perhaps state and local law enforcement officials, as well as FAA regulators, think that only the most problematic users are worth policing because they are the only ones who pose real harm to the public.

Meanwhile, in late 2018, several members of Congress accused Transportation Secretary Elaine Chao of ignoring a congressional directive to allow testing of self-driving vehicles only at a handful of government-designated proving grounds.[50] In the agency's autonomous vehicle guidance documents, the Department of Transportation (DOT) had taken a more flexible approach to allowing testing on roadways, thus avoiding the more restrictive approach recommended by Congress. Many states have followed the DOT's lead and essentially cobbled together guidance documents to keep things moving in the absence of more formal policy for autonomous vehicles.[51]

The FDA is also engaging in rule departure more regularly, although the agency has a long history of doing so using informal guidance documents.[52] Formal rulemaking procedures have always been challenged by the pace of technological change in the field of health and medical sciences. For better or worse, the FDA has used guidance documents and other notices to cope with the pacing problem in its policy area, and that pace now seems to be accelerating. In recent years, the FDA has issued guidance documents that outline best practices for conducting clinical trials,[53] "medical" smartphone apps,[54] and medical devices made through 3D printing,[55] among many other things.

These same soft-law trends are being used by many other agencies and will be discussed in greater detail in subsequent chapters. For purposes of our discussion here, it is enough to note that these are examples of government agencies ignoring their own laws or at least choosing to interpret and enforce them in a highly selective fashion. Philip Weiser and other scholars refer to that concept as *entrepreneurial administration*.[56] In essence, agency officials often act in a quasi-entrepreneurial way themselves by being rule-avoiders. Technically speaking, the underlying statutory powers of these agencies have not changed much in recent years. Practically speaking, however, the way the agencies choose to enforce their powers *has* changed in profoundly important ways.

It is unclear, for example, whether many of the new health and fitness smartphone applications we use today are perfectly legal under traditional FDA regulations. Likewise, many driverless car innovations are of ambiguous legality under federal, state, and local automobile regulations or traffic laws. And various 3D-printed products could be in violation of multiple bodies of law. But the amazing thing is that the agencies and officials overseeing these technologies are not in most cases trying to throw the book at innovators. In fact, most of these agencies are adapting in surprisingly flexible ways and allowing for a great deal of innovation to take place *even though they may be ignoring the enforcement of their own policies.*

Some sticklers for strong administrative law standards find such behavior quite troubling. That is understandable. When regulators ignore the law or bend it in extreme ways, it raises the prospect of unaccountable, nontransparent, extralegal government activity. On the one hand, that prospect is a serious problem for a nation built on respect for the rule of law and limited government. On the other hand, it is difficult to imagine that regulators will be able to keep up with the pace of change in high-tech markets and craft sensible policies unless they take a more flexible approach to policymaking. For those reasons, examples of rule departure will increase significantly in coming years, as will acts of evasive entrepreneurialism. The next chapter explores some of the reasons why we can expect those increases.

4

WHY EVASIVE ENTREPRENEURIALISM
IS ON THE RISE

Having considered various types of technologically enabled civil disobedience and their relation to traditional types of disobedience, we'll next explore some of the reasons people may be engaging in evasive entrepreneurialism more regularly than before. I will argue that several factors explain the rise in evasive activities—by both private parties and even some government officials themselves.

What Happens When Law Is Out of Touch with Reality?

As we've already seen, one reason we are told we should obey the law is because, well, *it's the law*! But sometimes laws and regulations can be just plain silly and defy common sense. Regulatory processes have grown complicated and incomprehensible, and the administrative state has become a sprawling behemoth, leading some scholars to conclude that "the modern administrative state is inconsistent with the rule of law."[1] Democratic processes are supposed to address such problems, but they often fail to do so.[2] The problems can lead to situations in which many people—including public officials themselves—begin to ignore the law altogether.

For example, there are century-old laws still on the books in some jurisdictions that require a person to wave red flags in front of a horseless carriage to warn others that it is coming down the street.[3] In fact, in Memphis, Tennessee, a woman is not supposed to drive a car unless a man is running or walking in front of her waving a red flag.[4] Meanwhile, in New Jersey, drivers are required by law to honk their horns before passing another vehicle.[5] "Of course, if you follow the law and honk your horn every time you pass, someone will probably shoot you," observes *Star Ledger* columnist Paul Mulshine.[6]

When Sony's Walkman portable music player became a national craze back in the early 1980s, communities in 10 states moved to ban its public use, citing amorphous public safety rationales.[7] For example, the town of Woodbridge, New Jersey, passed a ban on wearing portable headphones in public, which prompted a retiree named Oscar Gross to engage in civil disobedience by openly defying the law and daring police officers to arrest him.[8] Amazingly, the law is still on the books today.[9] Should people be arrested when they wear their headphones or earbuds while walking in Woodbridge?

More recent laws can sometimes be just as bizarre. For example, in 2011, Tennessee passed a law making it a crime to share your Netflix password too much, although the scope of this prohibition remains unclear.[10] And in 2013, Florida accidentally banned all computers and smartphones in the state while attempting to ban internet cafes, which was a misguided policy in its own right.[11]

Laws that seem to make sense in one era can become highly illogical in the next. For example, with the rise of fully driverless cars, humans won't always need to keep their hands on the steering wheel of vehicles. Yet state laws often mandate that drivers *always* have their hands on the wheel.[12] As noted earlier, the Federal Aviation Administration (FAA) has many rules on the books prohibiting drones from flying near airports, above other people, at night, or without first being licensed and registered with the agency. Yet the state and local law enforcement officials who are expected to document and report suspicious drone activity have shown neither the desire nor the requisite knowledge of those rules to enforce them. As of February 2018, no drone pilots have yet been fined for flying without a license, and only one drone pilot has received a warning.[13]

Must citizens obey such laws? One might be tempted to say that these are zombie laws that just never die, so who cares whether anyone pays attention? That's true but, again, some may answer: *it's the law!* And shouldn't we always obey the law? Apparently, many people don't really think so, because no one is paying attention to these laws, and even public officials turn a blind eye to their enforcement.

But what these examples hopefully prove is that just muttering "*it's the law*" is not a sufficient justification for expecting citizens to blindly swear allegiance to existing public policies. And quite often people don't. Instead, in these and a great many other cases, people go on living their lives and engaging in peaceful, voluntary interactions with others despite silly or nonsensical laws telling them not to do so. That disregard of existing laws may not be anyone's first-best solution, but obedience to a harmful law is potentially still worse.

Regulatory Accumulation and the Death of Common Sense

Furthermore, people sometimes ignore a law because there are so many laws that they can't understand them all, or they may not even know that a given law exists. Obviously, most regulations are put on the books with the best of intentions, and many rules *do* accomplish their intended objectives. But too much regulation can have a debilitating and destabilizing effect on an economy, a regulatory system, and democracy itself.

Philip K. Howard, chair of Common Good and the author of *The Death of Common Sense* and *The Rule of Nobody*, has written extensively about how regulatory accumulation has become a chronic problem. "Too much law," he argues, "can have similar effects as too little law." As he explains:

> People slow down, they become defensive, they don't initiate projects because they are surrounded by legal risks and bureaucratic hurdles. They tiptoe through the day looking over their shoulders rather than driving forward on the power of their instincts. Instead of trial and error, they focus on avoiding error.
>
> Modern America is the land of too much law. Like sediment in a harbor, law has steadily accumulated, mainly since the 1960s, until most productive activity requires slogging through a legal swamp. It's degenerative. Law is denser now than it was 10 years ago, and will be denser still in the next decade. This growing legal burden impedes economic growth.[14]

How much law are we talking about? For the past 25 years, Wayne Crews, vice president for policy at the Competitive Enterprise Institute, has documented the growth of federal regulation in his annual publication, *Ten Thousand Commandments: An Annual Snapshot*.[15] Since the first edition of his report was published in 1993, agencies have issued 101,380 rules.[16] "The estimate for regulatory compliance and economic effects of federal intervention is $1.9 trillion annually," Crews notes, which is equal to 10 percent of the U.S. gross domestic product for 2017.[17] When regulatory costs are added to federal spending, the burden equals $4.173 trillion, or 30 percent of the entire economy.[18] "If it were a country, U.S. regulation would be the world's eighth-largest economy, ranking behind India and ahead of Italy," Crews says.[19]

My colleagues at the Mercatus Center at George Mason University have meticulously documented how regulatory accumulation has had such deleterious effects on our lives and economy.[20] Patrick McLaughlin and Michael Wilt have pointed out that "the buildup of more and more regulatory restrictions distorts and deters the business investments that drive innovation and economic growth."[21] The net result of such an accumulation of red tape is fewer jobs, higher prices, and lower wages, they argue.[22]

A Deloitte survey of the U.S. Code reveals that 68 percent of federal regulations have never been updated and that 17 percent have only been updated once (see Figure 4.1).[23] Imagine if a business never updated its business model. It would likely eventually go under as the world around it changed. But that updating does not usually happen for regulatory agencies. The world changes, but they don't, and their rules don't—with serious consequences for both markets and the effective administration of government.

Mercatus research has shown that "[e]conomic growth in the United States has, on average, been slowed by 0.8 percent per year since 1980 owing to the cumulative effects of regulation," which means that the U.S. economy "would have been about 25 percent larger than it actually was as of 2012" if regulation had been held to roughly the same aggregate level it stood at in 1980.[24] These are

FIGURE 4.1

Number of times a section has ever been revised

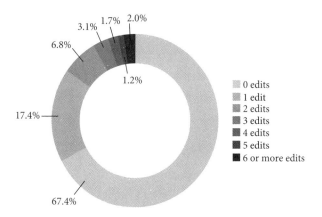

0 edits
1 edit
2 edits
3 edits
4 edits
5 edits
6 or more edits

Source: William D. Eggers and Mike Turley, "The Future of Regulation: Principles for Regulating Emerging Technologies," Deloitte Insights, June 19, 2018.

astonishing numbers that represent a significant loss of economic opportunities and benefits for society. "The more red tape and regulation, the less entrepreneurship and innovation," Crews concludes.[25]

Moreover, as Philip Howard suggests, at some point the "death of common sense" brought on by regulatory accumulation becomes a "good government" issue that raises profound questions for the health of our democratic republic. After all, can we really expect people to live up to the letter of the law when the law itself is increasingly incomprehensible?

And we're not talking about petty matters here. Many of these laws entail massive fines and jail time. Overcriminalization through federal criminal statutes has become such a problem that "regulatory offenses that purport to flesh out and refine the details of those statutes have proliferated to the point that, literally, nobody knows how many federal criminal regulations exist today," according to the Heritage Foundation.[26] Heritage has attempted to estimate that growing overcriminalization burden. "The number of criminal offenses in the U.S. Code increased from 3,000 in the early 1980s to 4,000 by 2000 and more than 4,450 by 2008," it finds.[27] Heritage also estimates "that nearly 5,000 federal statutes and more than 300,000 regulations contain potential criminal penalties."[28]

Worse yet, those numbers don't include the constantly growing body of laws and regulations at the state and local levels, where occupational licensing rules in particular create a litany of new crimes for noncompliance. Things have gotten so bad that it's now unsurprising when we hear of young kids being caught up in the dragnet of regulatory insanity, with permits even being required for children to run a corner lemonade stand in some towns.[29]

With no end in sight for such overregulation, technologies of freedom, evasive entrepreneurialism, and technological civil disobedience can help put beneficial pressure on policymakers by forcing them to reassess the wisdom of preexisting rules that no longer make sense. In Chapter 8, I will offer a few reform options that can help us clean up the messes we have created in the past. But first we will explore a couple other reasons why evasive entrepreneurialism and technological civil disobedience are increasingly common and explore how they might actually help solve chronic problems in government.

Risk-Averse Regulators and Build-and-Freeze Rules

It's not just that there's *too much* regulation on the books these days, but it's also the case that many of those rules are heavy-handed, inflexible, and slow to adapt to new circumstances.

Many agencies rely on a build-and-freeze model of regulation that puts rules in stone to deal with one set of issues one day, but then they fail either to eliminate the rules when they become obsolete or to reform the rules to bring them in line with new social, economic, and technical realities.[30] Cristie Ford notes that the problem with "old-style Welfare State regulation" is that it is "a clumsy, blunt instrument for achieving regulatory objectives" owing to its reliance upon "one-size-fits-all mandates, prohibitions, and penalties."[31]

One reason that policy rigidity, or "ossification,"[32] develops over time in traditional regulatory systems is that regulators tend to be highly averse to risk.[33] In his extensive work on bureaucracy, the political scientist James Q. Wilson observed that, to some extent, "[a]ll organizations by design are enemies of change," but that "government organizations are especially risk averse because they are caught up in a web of constraints so complex that any change is likely to rouse the ire of some important constituency."[34] That sort of risk-averse mentality views innovation as a threat rather than an opportunity. Conformity to process (i.e., doing things by the book) is valued much more highly than challenges to the status quo.

Risk aversion and status quo preservation tend to be especially prevalent among technocratic regulatory agencies that have the name of a particular technology in their titles. Historically, for example, regulators at the Federal Communications Commission (FCC) and the FAA have been quite hostile to new disruptive entrants and technologies that defied traditional classification. It is notable that both these agencies have historically relied to a high degree on the expertise of the very industry that they propose to regulate. They can hardly do otherwise, and yet in such cases the relationship between the regulator and the incumbent firms it regulates can become all too comfortable.

The Food and Drug Administration is another prime example of this sort of risk-averse culture in action. Henry Miller, MD, observed firsthand just how this problem manifests itself when he served as the founding director of the FDA's Office of Biotechnology. "A large part of regulators' self-interest lies in staying out of trouble. One way to do that," Miller writes, "is not to approve in record time products that might experience unanticipated problems."[35] But long delays create costs of their own, because, paradoxically, there is significant risk associated with taking no risks whatsoever. When important drugs or medical devices languish for years because of regulatory risk aversion, people are suffering. Some might even be dying. Jonathan Adler writes about how that type of delay played out in the 1980s with a drug to treat ulcers:

> The negative consequences of failing to quickly approve a lifesaving drug can be significant. Consider the example of Misoprostol, a drug that prevents gastric ulcers. Misoprostol was developed in the early 1980s and first approved in some nations in 1985. The FDA, however, did not approve Misoprostol until 1988. Even though the drug was already available in several dozen foreign countries, the FDA subjected Misoprostol to a nine-and-one-half-month review. At the time, between 10,000 and 20,000 people died of gastric ulcers per year. Had Misoprostol been approved more rapidly, it could have saved as many as 8,000 to 15,000 lives. Thus, in seeking to prevent one risk—the risk of approving an unsafe drug—the FDA contributed to the risk of gastric ulcers by preventing the use of a potentially lifesaving drug.[36]

Although misoprostol saves lives, and although it could have saved lives even before its approval, the drug met with considerable opposition because it is also capable of inducing abortion, a use for which no drug had been approved at the time. Activists urged a counterindication for pregnant women and a required pregnancy test before prescription, both of which were eventually added to the drug's treatment protocol. Yet both abortion and off-label prescription were entirely legal in 1988.[37]

Another recent example of FDA delays leading to potentially unnecessary deaths involved a product called MelaFind. The device allowed dermatologists to determine which moles are actually skin cancers. In his book *Innovation Breakdown: How the FDA and Wall Street Cripple Medical Advances*, Joseph V. Gulfo, MD, who helped develop the product, documented how the agency dragged the approval process on for years and let thousands of people die, he argues, because early detection can prevent death from skin cancers.[38]

Although the FDA's regulatory approval times are finally improving,[39] the problem Adler and Gulfo identify remains a specter that continues to haunt many innovators and emerging technologies (as we saw with the 23andMe case study earlier). Risk-averse regulators will not quickly think outside the box as new technologies invade the field of health care and medicine. Instead, they will more often simply force new technologies into old regulatory boxes, regardless of costs or other consequences. That reaction is even more likely to happen when powerful, entrenched incumbent companies are cheering on agencies and encouraging them to go slow—not really for public safety concerns but simply to keep out new competition.

Regardless of why it happens, when policymakers try to shoehorn new technologies into old regulatory regimes, the results are usually miserable. In the 1990s, for example, many governments across the globe viewed the internet as just another communications or media platform to be controlled, regulated, and taxed—just like telephones and broadcast systems were throughout the previous century. Fortunately, the United States made a fresh break with the failed policies of the analog era and let the digital economy flourish largely free of cumbersome layers of red tape.[40] The internet and digital technologies, at least here in the United States, are examples of sectors that were born free, as I discussed earlier. But many other technologies were born in captivity and were pigeonholed into old regulatory regimes.

Consider drones and driverless cars, which are both overseen by regulatory agencies housed within the Department of Transportation. In essence, the FAA and National Highway Transportation Safety Administration (NHTSA) believe their mission is to regulate anything that flies or rolls down the road on four wheels. Moreover, they both have a broad grant of power from Congress to comprehensively regulate in their respective fields.

So, when new technologies like drones and driverless cars come along, it is only natural that bureaucrats at the FAA and NHTSA would move to bring those technologies into the regulatory fold. That action is where problems begin, however. It is true that drones are flying things and that driverless cars are, well, cars. But these technologies are qualitatively different from their predecessors. The laws and regulations that applied to their industrial-era counterparts might not make any sense when applied to them.

For example, drones do not operate like planes or other aircraft, and they also do not carry the same risk of catastrophic harm when one of them crashes. That is not to say that drone safety isn't important and shouldn't be regulated perhaps, but it should be strange to think of drones and drone safety in the same way we do jumbo jets.[41] A different approach is needed that acknowledges the unique nature and circumstances of each technology.

The FAA has identified integrating drones into the American airspace as one of its top priorities, yet the agency insists on imposing its old regime of pilot licensing, aircraft registration, and tracking on commercial drone users and hobbyists alike. This course of action is consistent with the FAA's overly conservative and highly risk-averse culture, which was discussed earlier.[42] In 2018, FAA Acting Director Dan Elwell said the agency "need[s] assurances that any drone, any unmanned aircraft, operating in controlled airspace is identifiable and trackable."[43] Regulatory change is slowly coming to this sector, but the agency only seems willing to take incremental steps, and it looks to micromanage almost every aspect of drone innovation.

Likewise, old automobile safety regulations are so meticulously detailed that they specify minute requirements about the equipment in every car, right down to the steering wheel and pedals. But, to reiterate, we might not need any steering wheels or pedals in the robotic cars of the future. In fact, it might be better if such equipment were not in the cars of the future because humans might use them to interfere with the safe operation of truly autonomous vehicles, which will likely be able to drive us more safely than we can drive ourselves.

So, does it make any sense to mandate steering wheel or foot pedal safety rules for driverless cars? Believe it or not, there has been a heated debate going on about that issue in recent years. The wise thing for regulators to do would be to adopt a more flexible set of safety rules that did not try to technologically micromanage vehicle design but instead set broad safety goals and left it to others to figure out how to meet those goals in the most flexible way possible. In other words, regulators should adopt results-oriented performance objectives instead of rigid rule-driven design standards.[44] Regulation should be more *principles*-based as opposed to *process*-based.[45] Unfortunately, that's not the way top-down, build-and-freeze regulation usually works in practice.[46] By-the-book compliance is valued above all else, even when the rules are completely out of touch with new realities.

Archaic rules and regulations, as already noted, can be like zombies in that they are hard to kill and keep coming back from the dead. This longevity increases the likelihood that such rules are out of touch with new realities, especially in fast-moving technology sectors. "Nearly all federal auto safety regulations codify technical standards that are years, if not decades, out of date," says transportation policy expert Marc Scribner, a senior fellow at the Competitive Enterprise Institute.[47] He documents 73 federal vehicle safety regulations that incorporate 257 standards, "with half of them dating back before 1980."[48] These regulations have real-world safety effects. Scribner cites the example of adaptive driving beam headlights, which European and Japanese drivers already have access to. These headlights help prevent headlight glare that could temporarily blind drivers. Unfortunately, in the United States, archaic technical standards "require headlamps to have distinct high- and low-beam settings that make this safety-enhancing technology verboten" and "also [prohibit] the automatic engagement of emergency flashers, a technology necessary for self-driving cars to display hazard warnings to other road users."[49] Meanwhile, the NHTSA has pursued new technocratic mandates for vehicle-to-vehicle communications between connected cars, even though experts have documented how such premature standard-setting "locks in technology long beyond its usefulness."[50] Beyond adding unnecessary costs and potentially discouraging better standards from emerging in the long term, it is not even clear that the NHTSA's recommended standard would improve public safety in the short term.[51]

This example illustrates how traditional regulatory systems are often so top down and by the book in orientation that they become ossified and resistant to the common-sense adaptations that are needed to cope with technological change. Sometimes lawmakers or regulators may be so overwhelmed by the volume or complexity of their own rules that they are not aware of how they are thwarting innovative efforts.[52] Yet by freezing outmoded standards into place and then expecting slavish compliance with them, regulators often prohibit the sort of beneficial experiments that yield new and better ways of doing things.[53] Such rules focus on preemptive remedies that aim to predict the future and hypothetical problems that may not ever come about. Yet a better future cannot come about precisely because build-and-freeze regulations sometimes make it so difficult. Moreover, by forcing creators to seek special permissions to engage in innovative acts before they offer a new product or service, these types of regulations raise the cost of starting new businesses or launching new ventures, thus limiting new entry and competition. Economic studies find that "regulation has a negative effect on new firm creation and employment growth" and that large incumbent firms may benefit from using regulation to keep new rivals out of their hair.[54]

Excessive regulation can also block many other unseen activities that could benefit society.[55] Overregulation is particularly problematic when rules negatively affect technology-based startups, because empirical studies have documented how "these firms provide outsized contributions to employment, innovation, exports, and productivity growth."[56]

Innovators recognize these realities, of course. Regulations—especially illogical and inflexible ones—send entrepreneurs clear signals about what sort of behaviors public officials find acceptable. Policies like drone registration requirements simply incentivize entrepreneurs to behave in a more evasive fashion, relocate elsewhere, or abandon their plans entirely.

How Permits and Licenses Create Crony Capitalism

Evasive entrepreneurialism is also likely to occur when public policies favor incumbents and protect older companies and organizations from what economist Joseph Schumpeter famously described as the "perennial gales of creative destruction." These gales are what spur innovation and propel economies forward.[57]

When companies rest on their earlier achievements and seek to protect their turf or business models by making cronyism-based deals with policymakers, it is a sure-fire recipe for economic stagnation. First, as Tyler Cowen observes, "lobbying reorients the culture of a company toward politics and law, and away from innovation."[58] Second, and relatedly, economic research has consistently documented how "public rent-seeking can put a severe tax on innovative activities [and] hence sharply reduce the rate of economic growth."[59]

Unfortunately, political entrepreneurialism and seeking of privileges remain a chronic problem because, as Jeff Rowes of the Institute for Justice notes, "much economic regulation is rigged in favor of interest groups and against the public."[60] "Time and again, the losing interest groups created by scientific progress or technological change have been able to convince politicians to block, slow, or alter government support for scientific and technological progress," says Mark Zachary Taylor, author of *The Politics of Innovation.* "The losers and their political representatives have interfered with markets, public institutions and policies, and even the scientific debate itself—whatever they can do to protect their interests."[61] These acts of political favoritism "not only misallocate resources in the short term but they also discourage dynamism and growth over the long term," argue Brink Lindsey and Steven Teles in their recent book, *The Captured Economy: How the Powerful Enrich Themselves, Slow Down Growth, and Increase Inequality.*[62]

This is an old story, as we saw in Chapter 2 when discussing the research that Calestous Juma and other historians have done documenting industrial protectionism throughout the ages.[63] In the past century, the development of internal combustion and the rise of the automobile led to heated battles over the mechanization of agriculture and transportation, for example. "This was a gigantic Schumpeterian confrontation as the defenders of entrenched methods appealed to popular sentiments and tried to capture legal and political institutions to forestall the process of creative destruction," note two historians who have documented these skirmishes.[64] At one point in the 1920s, the Horse Association of America developed a sophisticated anti-motor lobbying effort to discourage tractors on farms and automobiles on city streets. It was doomed to fail, but the push for regulation resulted in initiatives in many cities to ban cars in favor of horses.[65]

Decades later, the story of automotive protectionism continues, except now it is the auto industry seeking regulatory advantages. Car dealers across America have pushed back against Tesla for having the audacity to try to sell its cars directly to consumers. Those car dealers do not like that Tesla's direct-selling business model would evade cronyism-based state laws, which require cars to be purchased through independent dealerships with exclusive service territories.[66] These laws restrict choice, raise prices for consumers,[67] and are really just about "protecting the dealer's privileged economic position as the middleman in the auto distribution chain."[68] It's no wonder that several states that passed laws protecting those dealerships from competition earned the ignoble distinction of receiving the 2014 Luddite of the Year Award, which is given annually by the Information Technology and Innovation Foundation to highlight "the worst neo-Luddite ideas that if followed would lead to reduced human progress."[69]

Many economic studies document how "incumbents, in particular, benefit from increasing levels of regulation" by using red tape to discourage entry and limit competition.[70] My Mercatus Center colleague Matthew Mitchell has done extensive work on how cronyism-based efforts such as these infect our political process, and he has developed a taxonomy of the "long list of privileges that governments occasionally bestow upon particular firms or particular industries." He notes the following:

> At various times and places, these privileges have included (among other things) monopoly status, favorable regulations, subsidies, bailouts, loan guarantees, targeted tax breaks, protection from foreign competition, and noncompetitive contracts. Whatever its guise, government-granted privilege is an extraordinarily destructive force. It misdirects resources, impedes genuine economic progress, breeds corruption, and undermines the legitimacy of both the government and the private sector.[71]

How and why does this problem persist? Scholars from various disciplines—economics, law, political science, history, and others—have explored the growth of what has been alternatively called the "interest group society,"[72] "receivership by regulation,"[73] "iron triangles," and "client politics."[74] What these concepts share is an insight that Mancur Olson identified in his 1965 book, *The Logic of Collective Action*. Namely, when benefits are concentrated and costs are dispersed (across all taxpayers, for example), interest groups form to take advantage of those benefits.[75] Those bearing the costs, which are diffuse and often hidden to some extent, will have less of an incentive to form groups to counter those receiving the benefits.[76] In that environment, government programs and regulations often become entrenched and self-perpetuating,[77] and political entrepreneurs exploit them and effectively "shape the political debate through agenda setting and the strategic use of cultural frames to tie individual interests with a vision of collective purpose."[78]

The history of communications and media policy in the United States throughout the 20th century provides a prime example of these problems. Under the auspices of serving the public interest, the FCC and many state and local officials repeatedly acted to repress innovation and competition.[79] The FCC effectively nationalized the wireless spectrum and imposed a highly politicized licensing scheme on all broadcast radio and television operations to strictly control entry and censor content disfavored by the unelected bureaucrats who ran the agency.[80] When unlicensed pirate radio operators tried to offer a little independent flavor in competition with licensed radio operators, the agency crushed most of those efforts.[81] And when in the 1950s a company named the DuMont Television Network looked to provide a nationwide alternative to existing TV broadcasters, the FCC effectively quashed those efforts and left American viewers without serious competition to the "Big 3" (ABC, CBS, and NBC) until the mid-1980s.[82]

After discouraging over-the-air competition, the FCC later acted to protect licensed TV broadcasters from new cable and satellite innovations using a variety of repressive techniques. At the same time, local governments were crafting cable television franchises that were "besmirched by scandal, political cronyism and vote-selling, illicit deals, and extortion."[83] The story was much the same for telephone networks, which witnessed decades of backward policies that limited competition at the federal, state, and local levels.[84] It was only with the rise of the internet and the digital revolution that citizens finally were able to get a taste of true competition for their eyes and ears, because online services were not subject to the same sort of special-interest rent-seeking activities.

Unfortunately, plenty of archaic rules and regulations remain on the books at the behest of special interests who rely on them for self-preservation.[85] State and local occupational licensing rules are particularly problematic in this regard, especially because such licensing regulations have expanded rapidly in recent decades. Economists estimate that roughly 20 percent of all professions are licensed nationwide (up from just 5 percent in the early 1950s), but licensing burdens vary widely by state.[86]

A large and growing body of economic research documents how occupational licensing restrictions result in higher prices and create barriers to new entry and innovation.[87] The Obama administration issued a major report on the costs of occupational licensing rules in 2015, documenting how the growing licensing burden results in a hidden tax on consumers of between 3 and 16 percent.[88] That report also documented how licensing rules limit economic mobility by "creating barriers to workers moving across State lines and inefficiencies for businesses and the economy as a whole."[89] In fact, research by economist Morris Kleiner and others has found that "restrictions from occupational licensing can result in up to 2.85 million fewer jobs nationwide, with an annual cost to consumers of $203 billion."[90]

These licensing regulations do not serve their intended benefit, because "most research does not find that licensing improves quality or public health and safety," the Obama administration concluded.[91] Even worse, the costs of licensing fall most heavily on immigrants and military workers, who need a new license to work each time they cross a state line.[92] In this way, licensing limits economic opportunity and upward mobility.[93] It particularly hurts the poor and disadvantaged, by making life hard for them both as workers and as consumers (by raising their costs or limiting the quality and quantity of goods and services available to them).[94]

Despite all these well-documented problems, occupational licensing rules keep expanding because they can be used strategically by incumbent interests.[95] In fact, state licensing boards for most professions are made up primarily of industry members who are already licensed and have an interest in keeping competitors out, as well as in keeping their own wages higher at the expense of the public, which then pays higher prices.[96]

It is unsurprising, therefore, that research reveals that "occupational licensing regimes, once enacted, almost never get revoked."[97] A 2015 report by the U.S. Bureau of Labor Statistics examined the de-licensing of occupations in the United States and found as follows:

> In nearly every instance that we analyzed, de-licensing and de-licensing attempts have been met not only with stiff resistance but also usually (when successful) with a movement to reinstitute licensing. Clearly, these results reflect the lobbying power of the occupations in question and their professional associations.[98]

This is the well-known problem of regulatory capture. Economists, historians, and political scientists have extensively documented this chronic, ongoing problem.[99] Few did a better job of highlighting this problem than the late economist Alfred Kahn, a self-described liberal Democrat, who was a major figure in the deregulatory efforts of the 1970s. As an academic, Kahn published a massive two-volume treatise, *The Economics of Regulation*, in 1970 that became a seminal textbook in the field. In it, he identified how capture was a particular problem for regulated industries:

> When a commission is responsible for the performance of an industry, it is under never completely escapable pressure

to protect the health of the companies it regulates, to assure a desirable performance by relying on those monopolistic chosen instruments and its own controls rather than on the unplanned and unplannable forces of competition. . . . Responsible for the continued provision and improvement of service, [the regulatory commission] comes increasingly and understandably to identify the interest of the public with that of the existing companies on whom it must rely to deliver goods.[100]

Kahn thought that regulatory capture was such a serious problem that, after President Jimmy Carter appointed him to serve as chairman of the Civil Aeronautics Board in 1977, Kahn promptly set out to dismantle the anti-consumer airline cartels sustained by government regulation and then abolish the agency altogether.[101] Airline routes and options expanded rapidly thanks to the efforts of Kahn and President Carter to stop this cartel-type arrangement.

Cronyism and capture continue to be serious problems.[102] "Not one word of law can be changed without a majority of Congress running a gauntlet of special interest influence," Philip Howard notes. "There's a reason Congress doesn't even consider fixing old laws: It's unthinkably difficult," he says.[103] The problem is even worse in some states and cities. As the Tenth Circuit Court of Appeals memorably observed in one 2004 decision, "while baseball may be the national pastime of the citizenry, dishing out special economic benefits to certain in-state industries remains the favored pastime of state and local governments."[104]

Cronyism and capture tilt the political playing field violently in favor of old interests, old ideas, and old business models.[105] This tilting makes it difficult if not impossible to achieve positive reform by the book or through the usual processes; those books and processes are largely controlled by vested interests. By extension, it should not be surprising that some innovators will seek to do an end-run around the process in an attempt to offer the public new and better alternatives to the crony capitalist status quo. Later, in Chapter 8, a wide variety of licensing reform options are discussed.

Demosclerosis and Kludgeocracy

The problems identified here—illogical laws, regulatory accumulation, inflexible build-and-freeze regulatory regimes, and cronyism and capture—should all be ripe reform opportunities for elected lawmakers in Congress or state legislatures. After all, that type of action by lawmakers is what representative democracy is all about; our leaders are supposed to be accountable to the people and the constitutional principles that we cherish. When things are not working properly, lawmakers should use stepped-up oversight and sensible reforms to make sure government remains accountable and effective.

Alas, that rarely happens anymore. Little has changed since Hannah Arendt wrote the following in 1972:

> Representative government itself is in crisis today, partly because it has lost, in the course of time, all institutions that permitted the citizens' actual participation, and partly because it is now gravely affected by the disease from which the party system suffers: bureaucratization and the two parties' tendency to represent nobody except the party machines.[106]

In such an environment, it should not be any surprise that rules accumulate without any care in the world as to whether they make sense and that agencies are allowed to run wild and impose new constraints on our liberties without anyone holding them accountable for it. "Although these bureaucracies are theoretically part of the executive branch, presidents cannot effectively control them," observes Peter J. Wallison. "Congress also seems powerless to control their growth."[107] This change has happened because Congress has abrogated its constitutional responsibilities, casually delegating much of its legislative power to regulatory agencies. This development has made Congress an increasingly ceremonial branch of government and has given agencies massive control over important decisions that are supposed to be made by the democratically elected leaders.[108] Unsurprisingly, as noted, powerful interests tend to control those regulatory processes. The consent of the governed lacks much meaning in a system where special interests get heard while innovators and average Americans are widely ignored.

In his 1999 book, *Government's End: Why Washington Stopped Working*, Jonathan Rauch coined the term "demosclerosis" to describe "government's progressive loss of the ability to adapt."[109] "[A]s layer is dropped upon layer," he argued, "the accumulated mass becomes gradually less rational and less flexible."[110] Steven Teles coined the term "kludgeocracy" to describe the way that government solutions frequently are clumsily cobbled together to patch past problems and create temporary fixes.[111] "The complexity and incoherence of our government often make it difficult for us to understand just what that government is doing," Teles says.[112] Kludgeocracy creates serious costs for individual citizens, their governments, and our democracy, he argues.

Demosclerosis and kludgeocracy are the new normal in the United States today. Even those working inside the legislative branch agree that government has largely lost its ability to adapt to technological changes. An August 2017 survey by the

Congressional Management Foundation "found overwhelming majorities of senior congressional aides believe Congress is not equipped to execute its basic functions."[113] The most cited areas of concern by congressional staff dealt with the lack of both the skills and abilities as well as adequate time and resources "to understand, consider and deliberate policy and legislation."[114] As Congress has lost its expertise and understanding, the regulatory bureaucracy has continued to grow.[115] In turn, this growth of bureaucracy has fueled the cronyism and capture problem discussed previously.[116]

There are no easy fixes for these problems, but the best option is to use "simple rules for a complex world," to borrow a phrase from legal scholar Richard Epstein.[117] Given the current state of modern legislative politics and policymaking for emerging tech in particular, such reforms will be challenging. As noted, one useful step in that direction would be to shift away from top-down, command-and-control rules and regulations and toward bottom-up, flexible, principles-based rules.[118] We also need reforms that are more comprehensive and aimed at clearing away the accumulated regulatory deadwood.

If policymakers fail to adapt regulatory systems to new realities, the ongoing evolution of technology and the evasive efforts of entrepreneurs may represent our last, best hope for correcting the problem of rules being so out of step with reality and common sense. Only by keeping government officials on their toes and forcing them to reevaluate or abandon misguided, outdated policies will we at least have some hope of restoring serious checks and balances within our system.[119]

At a minimum, it should be clear that the growing problems of demosclerosis and kludgeocracy are likely to be major contributing factors fueling the growth of evasive acts. When entrepreneurs cannot understand the complexities of a regulatory system that defies common sense, it should not be surprising that they look to evade it altogether.

The Pacing Problem

A final reason that evasive entrepreneurialism and technological civil disobedience will occur with greater regularity is that the pacing problem has become an undeniable feature of modern life.[120] As already discussed, the pacing problem refers to the fact that the pace of technological change seems to be constantly racing ahead of law's ability to keep up with it.

"There has always been a pacing problem," argues Yale University bioethicist Wendell Wallach, author of *A Dangerous Master: How to Keep Technology from Slipping beyond Our Control*.[121] But he and many other science and technology experts believe that innovation today is unfolding at an unprecedented clip, making it harder than ever to govern emerging technologies using traditional legal mechanisms.[122] "The faster the rate of change, the more difficult it becomes to effectively monitor and regulate emerging technologies," Wallach claims. "Indeed, as the pace of technological development quickens, legal and ethical mechanisms for their oversight are bogging down."[123] Technology policy scholar Larry Downes also refers to this idea as the "law of disruption," or the fact that "technology changes exponentially, but social, economic, and legal systems change incrementally."[124]

Even government officials acknowledge the challenge presented by the pacing problem and the so-called law of disruption. In a 2016 speech about private drones, then-FAA administrator Michael Huerta noted, "I have said more than once that innovation moves at the speed of imagination and that government has traditionally moved at, well, the speed of government."[125] A few months after Huerta made those remarks, the Department of Transportation released a report on the regulation of driverless car technology and noted, "The speed with which [driverless cars] are advancing, combined with the complexity and novelty of these innovations, threatens to outpace the Agency's conventional regulatory processes and capabilities."[126]

Similarly, FDA regulators have increasingly referred to the pacing problem when discussing the challenge of keeping up with new medical innovations. As noted earlier, the agency has been trying to crack down on what it regards as rogue stem cell treatment clinics. But it is proving to be quite a challenge. "There are hundreds and hundreds of these clinics," says Peter Marks, MD, director of the FDA's center for biologics evaluation and research. "We simply don't have the bandwidth to go after all of them at once."[127] The challenge for the FDA and other health regulators is only going to intensify as technological capabilities expand and costs fall. For example, according to the National Human Genome Research Institute, the cost associated with DNA sequencing is already falling at a faster rate than even Moore's Law (Figure 4.2).[128] That principle is named after Intel cofounder Gordon E. Moore, who first observed that the processing power of computers doubled roughly every 18 to 24 months while the prices of computers remained fairly constant. Moore's Law has been a relentless force in information technology markets, forcing firms to rethink their business models constantly to stay on the competitive edge. It will be hard for regulators to slow the pace of genetic innovation if those trends continue and more people can gain access to those technological capabilities.

Why is the pacing problem accelerating? Mainly because the technologies of freedom discussed throughout this book work in a symbiotic fashion. Gary Marchant writes of "concurrent technological revolutions" taking place.[129] These technologies build

FIGURE 4.2
Cost per genome

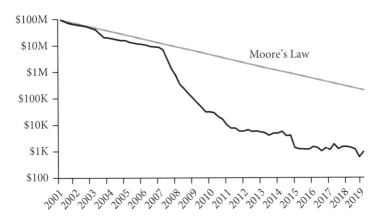

Source: National Human Genome Research Institute, "DNA Sequencing Costs: Data," accessed January 24, 2019, https://www.genome.gov/27541954/dna-sequencing-costs-data/.

Note: K = thousand; M = million.

upon each other and grow more powerful as their shared foundational elements—microchips, sensors, wireless networking and geolocation capabilities, digital code, cloud computing, cryptography and anonymization tools, and more—become faster, cheaper, more powerful, and easier to find and use.[130] Many of these capabilities give rise to what regulatory scholar Cristie Ford refers to as "seismic innovations" that completely catch the world and regulators by surprise. As a result, they completely alter markets and legal ecosystems.[131] As entrepreneurs continuously find new ways to combine these technologies, and as citizens come up with creative new ways to apply them in many different fields, their disruptive potential continues to expand. In turn, the pacing problem grows across many sectors.[132]

Consider what is set to happen in the field of health care, thanks to those technologies working together. "Pervasive computing technologies, such as blockchain, will keep health care data decentralized and closer to the patients," notes a 2018 report in the *MIT Sloan Management Review*. The report also notes:

> Other relevant data will be kept in blocks close to health care providers. Machine learning can be added to the blockchain to break down the traditional sequence of clinical trials and treatment. Patients, researchers, pharmaceutical companies, and physicians can collectively use this data to run trials and realize benefits that are customized and available more quickly. When 3D printing capability for drugs and medical devices is added to this smart chain, a whole new form of health care technology will be born, one that radically changes the traditional chain of treatment and prevention.[133]

Every sector of our economy is poised for this sort of disruption as "software eats the world"[134] and new technologies of freedom continue to be developed and combined to address every human need and want under the sun.

But radical innovation also gives rise to radical uncertainty—for markets and especially for governmental actors. As discussed, top-down legal and regulatory mechanisms have always been unwieldy and slow to adapt to new technologies and circumstances. But this problem has become far more pervasive in technological governance efforts.[135] "Twentieth century regulatory institutions are ill-equipped to effectively address the rapid progress driving twenty-first century technologies," observes Ryan Hagemann.[136] Even when laws and regulations are proposed to address new technologies, they are often outmoded before they are finalized. That's probably one reason we seem to see fewer technology-oriented laws and regulations achieving final implementation today. When it takes years to advance formal rules and regulations through legislatures and administrative agencies, it is less likely that policymakers will be able to get new policies in place in a way that is timely and effective. In turn, this slowness opens the door to still more evasive entrepreneurialism because technology innovators and users will know that laws likely won't be able to keep up with all the potential uses of new tools and capabilities.

Synthesis: Why Now?

To recap, this chapter has asked what happens in a world in which the following phenomena are at work:

- Laws and regulations are increasingly out of touch with common sense.
- Regulatory accumulation continues unabated.
- Build-and-freeze regulation stifles efforts by agencies to innovate.
- Cronyism and capture continue to be chronic problems.
- Demosclerosis is the new normal as elected leaders refuse to do anything to reform broken systems or adapt government policies to new realities.
- The pacing problem is accelerating and making it harder than ever for public policy to keep pace with the technological changes we are witnessing.

This is a world in which the potential for direct and indirect forms of evasive entrepreneurialism and technological civil disobedience will likely expand over time. In this environment, evasive entrepreneurs will increasingly operate at the margins of the law to push new innovations to market. I've already explored many reasons why that type of activity can have benefits for society and our economy. But next I will discuss what may be the most beneficial byproduct of evasive entrepreneurialism and technological civil disobedience: it can ensure that our laws and lawmakers are more accountable and transparent, that technology policy is rooted in common sense, and that important civil and economic liberties are protected.

5

INNOVATION AS CHECKS AND BALANCES

The previous chapters explained why evasive entrepreneurialism is on the rise, offered some prominent examples of it in practice, and discussed how these developments are affecting technological governance efforts. This chapter briefly explores the interplay between innovation and government power. Specifically, it considers how evasive entrepreneurialism and the freedom to innovate can play an important role in checking the power of the state and can help protect important liberties.

I will argue that the relationship between technology and the scope of government power is not always obvious but that innovative activities can play an important role in checking some of government's worst tendencies *at the margin*. Citizens are increasingly using innovations and new technological capabilities to push back against authorities who have lost touch with common sense or failed to adapt public policies to the will of the people.

Innovation gives people additional voice and exit options to protest and push back against broken political systems or, more simply, to just help them live lives of their own choosing. Importantly, this notion of innovation as a form of checks and balances has the additional benefit of being a more reasonable way to reclaim some of our personal sovereignty, at least compared with more radical steps. I will argue that, in effect, innovative acts are like mini-revolutions that help us bring about positive change through small acts of technologically enabled creativity and resistance.

The Failure of Recent Reforms to Constrain Excessive Regulation

The previous chapter reached some dismal conclusions. Most notably, special interests have increasingly captured government processes and institutions. Meanwhile, the sprawling administrative state continues to grow without much constraint. Finally, demosclerosis and kludgeocracy have become such chronic diseases within our body politic that sensible reforms seem more elusive than ever.

Over the past few decades, reformers have advanced a wide variety of measures aimed at constraining the worst impulses and excesses of the administrative state, especially at the federal level. Those reforms have included cost-benefit analysis, regulatory budgeting, agency reorganization, and downsizing. Some of those reform ideas have been put into place, but those efforts have largely failed to constrain the regulatory state in any meaningful way or even to help clean up outdated, inefficient, and illogical laws and rules. As documented earlier, the federal regulatory burden has expanded constantly over the past quarter century. That has held true regardless of which party was in power in Congress or the White House.

Meanwhile, deregulation and agency reform have stalled. In an era in which trust in government is at an all-time low and government failures are multiplying, one would think that at least a couple of agencies would have been put on the chopping block for failing to do their job effectively. But that almost never happens. Most people probably could not name the last time a federal agency was significantly downsized or abolished. The Civil Aeronautics Board, which was abolished in 1985, and Interstate Commerce Commission, which was abolished at the end of 1995, are two notable but rare examples of agencies that outlived their usefulness and were terminated. The closures occurred after academics and policymakers documented how those agencies failed miserably to serve the public interest, partially owing to the problem of regulatory capture.

Meanwhile, the constitutional checks and balances that were meant to constrain the power of the executive and legislative branches of government have been gradually eroded or simply ignored.[1] What about the state governments and their role in checking federal power, which was a core feature of the original constitutional order?[2] States were supposed to retain significant powers of their own that could help constrain the federal government through a system known as federalism. This system was also designed so that states could act as a sort of relief valve that would allow citizens to "shop around" for better jurisdictional governance arrangements.

Alas, America's federalist system has not helped constrain the growth of the federal government much either. Perhaps the most intense competition among the states today is the battle to see who can get more federal money, much of it to help administer federally mandated entitlement programs. Worse yet, most states have also allowed their internal administrative regulatory systems to run amok and fostered anti-competitive, anti-innovation laws and regulations—especially excessive licensing regimes—that have limited new entry and innovation at lower levels of government.[3] Taken together, as Steven Teles observes, "there are reasons to question the idea that federalism and the separation of powers limit the growth of government" because "the activity of the American state is not significantly more limited than those of other industrialized countries."[4]

On the one hand, the judicial branch, compared to Congress, has been a bit more effective in checking the power of the administrative state and reining in regulatory excess at both the federal and state levels.[5] On the other hand, the courts have simultaneously been remarkably deferential to regulatory agencies and have done little to actually regulate the regulators in any meaningful way.[6] By their nature, courts can only interpret the law as written, not craft new law, which makes Congress's abdication of lawmaking to the executive branch all the worse. Moreover, even when the judiciary does check overreach by other branches, those victories sometimes take years or even decades to wind through the courts, or they may only result in incremental constraints.

This situation could change, but for now it seems reasonable to conclude that traditional legal and judicial efforts aimed at checking the power of regulatory bodies have been, and will continue to be, largely ineffective.

Does Technology Constrain Government Power or Expand It?

Before we consider the claim that innovation and evasive entrepreneurialism can help check state power and improve government processes, an important caveat is in order. Technology is not some sort of magic pixie dust that can be sprinkled on governments to make them smaller, smarter, or more accountable.

The mere existence of the new technologies and capabilities discussed throughout this book will not instantly make governments more responsible or responsive to the people.

The case presented here is more of a *marginal* one, not a grandiose utopian theory of political liberation through technological change. Technology can constrain governments *at the margin* in important ways. But it takes hard work and effort to successfully constrain state power in any meaningful way. Combined with sensible policy efforts, the right combination of creative minds, innovative technologies, and wise policies can help keep governments constrained and accountable.

Technology can sometimes also empower governments and make them even more repressive, however. At a macro level, consider what history teaches us about the relationship between technological innovations and the growth of the state. To the extent that there is a clear relationship, many libertarian-minded analysts may not care to acknowledge it. My Mercatus Center colleague Tyler Cowen has pointed out that modern technology (advanced transportation and communications networks in particular) greatly facilitated the growth of government in the 20th century.[7] These technologies can make individuals and institutions more detectable and traceable, and therefore more easily regulated. "Future technologies may either increase or decrease the role of government in society," he notes, "but if history shows one thing, it is that we should not neglect technology in understanding the shift from an old political equilibrium to a new one."[8]

Cowen's warning is wise. Technology can be used for good and bad ends, and it has been used (and abused) by governments to expand their powers and limit our liberties.[9] "The natural progress of things is for liberty to yield, and government to gain ground," Thomas Jefferson correctly observed during the founding of our constitutional system of government. It is hardly surprising that governments would take advantage of new technological capabilities to do exactly what Jefferson protested in America's *Declaration of Independence*: "erect[] a multitude of New Offices, and [send] hither swarms of Officers to harass our people and eat out their substance." Jefferson would have been shocked to see the extent to which America's federal government uses modern technologies to spread its reach today.

Although technology's role in helping to facilitate the growth of government during the past century is undeniable, it also would be incorrect to conclude that it has been unilateral or without corresponding tradeoffs for governments or others in positions of power. Some technologies have facilitated centralization and control, but many other modern technologies have empowered citizens in new and unique ways and have often helped them check the power of governments and other powerful agents. Engineering historian Samuel Florman argues that it is important to put things in even broader historical perspective when one evaluates the relationship between innovation and power. He says that "it simply is not true that advances in technology have been helpful to the Establishment in increasing its power over the masses. In fact, the evidence is all the other way. In technologically advanced societies there is more freedom for the average citizen than there was in earlier ages."[10]

Again, what is interesting is how and in what specific contexts new technologies expand the potential for evasive entrepreneurialism and permissionless innovation at the margin.

Recall the high hopes some early internet pundits had in the 1990s for the potential liberating power of the internet and encryption technology, and compare them to the way things actually unfolded. As noted, governments found ways to regulate many online activities and continue to engage in widespread surveillance despite improvements in encryption technologies.

Today some critics look back and snicker at early techno-utopian predictions that the digital revolution represented an emancipatory moment that would free people from overbearing governments. Tech critics are fond of derisively pointing to John Perry Barlow's impassioned 1996 manifesto, "A Declaration of the Independence of Cyberspace."[11] In the widely cited polemic, Barlow offered up a highly utopian and deterministic view of the emerging digital world: Not only do governments have no business messing with cyberspace, he insisted, but also the nature of the internet and how it is technically constituted make control of information impossible. He argued that, as a result, "the global social space we are building [will] be naturally independent of the tyrannies you seek to impose on us."[12]

Barlow was hardly the only person espousing such views at the time.[13] Other notable early internet exceptionalists included Nicholas Negroponte, the former director of the MIT Media Lab, and George Gilder, the prolific technology author and visionary.[14] Although neither went as far as Barlow, both Negroponte and Gilder wrote passionately about the empowering nature of digital technologies and the infeasibility of government regulating them effectively. In his 1995 book, *Being Digital*, Negroponte argued that "[l]ike a force of nature, the digital age cannot be denied or stopped."[15] Similarly, Gilder's books and essays inspired the early generation of Silicon Valley enthusiasts who tended to, in the words of critic Thomas P. Hughes, "dismiss the past as irrelevant and believe that their world is entirely new under the sun. They fervently believe that computer-driven technology will change everything."[16]

Things did not turn out the way that Barlow and other Net evangelists hoped or predicted. As various internet realists would point out in critical articles[17] and books,[18] the internet exceptionalists went much too far in insisting that these new technologies would suddenly liberate society from oppressive government policies or at least be immune to government overreach.

Will Technology Erode the Foundations of the Nation-State?

Many early internet exceptionalists went too far with their salvation-through-technology enthusiasm, but the internet realists went too far in their own countercriticisms. The internet and digital tools and platforms really *did* disrupt the ability of governments to exercise some of their traditional powers. It is essentially game over for the sort of censorship we witnessed in the analog era. Here in the United States, for example, there are still laws on the books regarding indecency or obscenity, but regulators at the Federal Communications Commission, the Department of Justice, and other bodies have largely given up on trying to enforce them.

It is also harder now (but not impossible) for nation-states to control public opinion, dissent, and protests. In fact, many of the same realist scholars who reject internet exceptionalist thinking have nonetheless argued that digital technologies have become a remarkably destabilizing force for nation-states or are undermining democracy and the rule of law.[19] In his 2015 book, *Disruptive Power: The Crisis of the State in the Digital Age*, Taylor Owen argues that

> the state is losing its status as the pre-eminent mechanism for collective action. Where it used to be that the state had a virtual monopoly on the ability to shape the behavior of large numbers of people, this is no longer the case. Enabled by digital technology, disruptive innovators are now able to influence the behavior of large numbers of people without many of the societal constraints that have developed around state action.[20]

Many others have made similar arguments and pointed out that when one examines the long arc of history and explores how technological change influences power, innovations (and new information technologies in particular) "very often decentralize power, and this fact is not graciously conceded by the powers that be."[21] Cowen's earlier assessment is certainly correct regarding the way some modern technologies facilitated the centralization of government powers during the past century. But it may be the case that centralization was a historical anomaly, and now the aggregate effect of so many combined technological innovations is overwhelming the capabilities of governments and traditional governance systems. Some scholars even wonder if we are witnessing the end of the Westphalian era and the modern nation-state because, as Stephen J. Kobrin has written, "[t]he clear line between domestic and international transactions is again becoming ambiguous and blurred."[22] "With the emergence of an integrated global economy," he argues, "it is increasingly difficult to determine what is a national product, a national technology or even a national firm."[23]

That statement is particularly true of global information technology companies with the rise of the internet, smartphones, and social media platforms. In a sense, Barlow and those much-ridiculed internet evangelists of the 1990s have been partially vindicated as today's leading tech critics proclaim that large multinational companies are essentially becoming their own sovereigns (sarcastically labeling them things like "Facebookistan" and "Googledom") and supplanting, or at least undermining, the power of nation-states.[24] Since the dawn of the internet, debates have raged over the question of whether cyberspace was a place at all and the degree to which the set of technologies and platforms that constituted that place presented a serious challenge to traditional hegemons.[25] There was no real winner in those academic deliberations but instead a begrudging acceptance on behalf of most scholars on both sides that, as David Post concluded, "[t]he internet surely complicates statehood and statecraft, but hardly renders either irrelevant."[26]

One thing should be clear, however. Although Facebookistan and Googledom are cute labels for large and powerful companies, private companies have no true sovereignty over us—at least not in the traditional meaning of the term.[27] No matter how big they are, information platforms like Facebook and Google do not possess the coercive powers held by actual nation-states. People who use such private platforms are free to escape their territories any time they want or to not even join them in the first place.[28] Moreover, we should not forget the many fears and apocalyptic predictions we heard just a generation ago about earlier corporate sovereigns like AOL and Microsoft, which were toppled off their lofty perches after Schumpeterian waves of creative destruction decimated their hegemony in fairly short order.[29] That disruption is simply not something that happens easily or very often with nation-states.

Technology's Countervailing Effects

On the other hand, the growing multitude of private innovators and technology providers *can* supply us with new ways of expressing our opinions about, and taking actions to challenge, the power of nation-states, their laws, and their leaders. Clay Shirky,[30] Yochai Benkler,[31] and Martin Gurri[32] have explored the disintermediating role of the internet and digital networks in recent years. These authors argue that the falling cost of information production and the ease of digital distribution make it increasingly easy for individuals to engage in group-forming and collective action. The combination of these new technologies and capabilities can be highly disruptive to well-established organizations and authorities. "The information technologies of the twenty-first century have enabled the public, composed of amateurs, people from nowhere, to break the power of the political hierarchies of the industrial age," argues Gurri.[33] This "revolution in the nature and content of communication," he says, "has ended the top-down control elites exerted on the public during the industrial age,"[34] so much so that it now poses "an existential challenge to the legitimacy of every government."[35]

One notable example of this challenge of authority was the Arab Spring uprisings that developed through northern Africa and various Middle Eastern countries beginning in the spring of 2011. Smartphones, social media platforms, and video distribution sites proved too hard for governments to control, and news of the protests spread rapidly by the efforts of ordinary citizens. Some governments fell as a result.

At times, it may seem more like a wash, however, with new technological developments empowering *both* the state and the citizenry in different ways. Although digital media and communications platforms empowered pro-democracy Arab Spring protest movements, they also empowered extremist religious movements, terrorist groups, and other repressive governments.[36] Gurri points out that the technologically enabled "revolt of the public" can be a mixed bag in this regard—we celebrate how it facilitates protests against authoritarian regimes but decry it in the case of ugly forms of populism rising up in the United States and Europe.[37]

Similarly, what are we to make of the state secrets revealed by WikiLeaks and Edward Snowden over the past decade? To an unprecedented degree, state secrets are leaking out faster than ever because whistleblowers have been able to use digital technologies to obtain and make public government information that would never have seen the light of day in the past. Yet from those revelations we have also learned that to an equally unprecedented degree, governments are using new digital capabilities to surveil the citizenry and vacuum up mountains of data about both domestic and foreign individuals.[38]

During the "crypto wars" of the late 1990s, many internet advocates and scholars (admittedly including me) believed that encryption technologies would make it much harder for government snoopers to pry into our business and violate our privacy. To some extent, that theory was correct. The more people and organizations encrypted data transmissions, the harder governments had to work to surveil communications and data flows. In true *Empire Strikes Back* fashion, however, many government law enforcement agencies simply adapted and concocted backdoor solutions to circumvent technologies of resistance.[39] Worse yet, government bodies strong-armed companies and forced them to quietly hand over mountains of data about us.[40] What are we to make of this? Are we better off because we will live in a more transparent society and have access to more state secrets than ever before? Or are we worse off because the potential for state surveillance has expanded significantly because of digital collection and retention capabilities?

Some will insist that, when we think about privacy in the abstract, we have witnessed a net loss of freedom owing to the rise of the internet and digital technologies.[41] But that net loss is questionable when one considers the many underappreciated ways in which digital technology has actually *expanded* privacy in recent years.

Consider, for example, young people doing a search for information online about sexuality or contraceptive issues that they are still struggling to understand or are frightened to ask someone else about. Similarly, consider adults who, before the internet, had to walk into an adult bookstore or video rental store to purchase controversial material but who now can do so in the confines of their own home, free from the unwanted gaze of others. One can debate the morality of these choices, but two things seem clear. First, digital technologies greatly expanded the volume and accessibility of various types of sensitive or controversial information that many people have likely long desired. Second, digital systems also made it easier to find and consume such information without needing to deal with the awkwardness of potentially overbearing figures or authorities. From the perspective of personal privacy rights, therefore, this change seems to be a major win for individual freedom. Benjamin Wittes and Jodie C. Liu make that point in a 2015 Brookings Institution white paper on "privacy paradoxes." They write:

> Many new technologies whose privacy impacts we fear as a society actually bring great privacy boons to users, as well as significant costs. Society tends to pocket these benefits without much thought, while carefully tallying and wringing its hands about the costs. The result is a ledger in which we worry obsessively about the possibility that users' internet searches can be tracked, without considering the privacy benefits that accrue to users because of the underlying ability in the first instance to acquire sensitive material without facing another human, without asking permission, and without being judged by the people around us.[42]

Of course, people weigh privacy wins and losses differently, so it is hard to evaluate what those privacy issues mean for society as a whole.[43] The broader point of these examples is that technology sometimes does empower governments while also empowering individuals to push back against governments in new and different ways, thus challenging authority through technologically enabled dissent and even disobedience.

This sort of tit-for-tat process will continue to play out in coming years. This development transcends information technology, too. Jason Kuznicki, author of *Technology and the End of Authority*, posits that we are witnessing the rise of what he calls "technocratic de-government" through "technologies, both physical and social, that will allow us more often to do without the coercion that is inherent to governing and thus to the state."[44] The growing multiplicity of technological tools and capabilities, he argues, poses "an ongoing challenge to political authority."[45]

For example, the rise of Bitcoin and other cryptocurrencies is already prompting the same questions encryption technologies did a generation ago. Chapter 2 discussed how, as Cowen argues, "a blockchain is actually a form of governance and that is what makes it such a potentially radical idea."[46] Governments are already struggling to adapt to these new realities. Some libertarian theorists like to suggest that Bitcoin, cryptocurrencies, and blockchain technologies pose a sort of existential risk to the nation-state.[47] We have heard that argument before, as noted, and government has proved more resilient and adaptive than many originally had anticipated. Nonetheless, today's cryptocurrencies, decentralized marketplaces, and darknets appear to already be challenging many traditional powers that nation-states have long taken for granted, including the power to control monetary transactions and the exchange of various types of goods across their borders.

It is important to take a holistic view of the growing constellation of technologies of freedom and resistance when considering how technologies influence state power. The myopic focus on encryption in the past and blockchain today presented these technologies almost as silver bullets that could swoop in to save the day. We should reject monocausal narratives about how any particular technology will challenge and constrain government.

We should instead embrace a theory and an approach that look at the vibrant and growing class of technologies that can, *in the aggregate and at the margin*, provide a positive check on government power.

Going forward, if the theory that modern technology really is becoming increasingly difficult to control and if average citizens will use those technological capabilities to skirt the law and live lives of their own choosing, then we'll be forced to think hard about how society and governance institutions will change. To reiterate, it would be ludicrous to suggest that nation-states are going to wither away because of the onslaught of emerging technologies and evasive entrepreneurialism. Yet it would be equally foolish to suggest that those developments would not have far-reaching ramifications.[48]

Many supporters of expanded regulation admit that technology is challenging government power, but they cast this fact as a problem to be solved rather than an opportunity to be embraced. In her recent book *Innovation and the State*, financial regulation expert Cristie Ford uses water metaphors to describe the way that "innovation runs down avenues of opportunity" and "will run under, around, and over obstacles (including seemingly clear regulatory requirements)."[49] Thus, she presents "innovation as a key challenge to regulation itself,"[50] treating it as an almost existential threat to the goals and effectiveness of the regulatory state. The other way to look at it, as I suggested in Chapter 3, is that innovation and acts of evasive entrepreneurialism represent a form of dissent that challenges the legitimacy of regulatory systems that have, for whatever reason, become inflexible, unaccountable, and just out of step with modern needs and realities.

Whether people agree with the thesis I have set forth here, they should be willing to explore the profound questions these developments raise for the future of nation-states, their legitimacy, and traditional governance mechanisms, as follows:

- Will government policies and institutions lose even more of their legitimacy if citizens use new technologies of freedom to evade or ignore a growing array of state edicts?
- What constitutes the consent of the governed when a large portion of a society is essentially using new technology to opt out of traditional governance mechanisms?
- Do alternative governance mechanisms or other solutions exist to deal with the potential downsides of increasingly unregulated technological innovations?

The uniquely American contribution to this discussion flows from the radical idea that inspired the American Revolution: sovereignty lies in the people themselves. Only when the people delegate some of that power to the state does government come to have any legitimacy whatsoever. Through the consent of the governed and the requisite constitutions or other contracting elements, "we the people" transfer power to governments to handle a variety of things that we deem better not left to private actors or actions. An important corollary is that, because true sovereignty lies with the people, even after we transfer some of it to the state, we can later change our minds and take that power back. Unfortunately, reclaiming authority is much easier said than done, especially if we are talking about the radical steps sometimes required to accomplish the return of that power to the people.

What I have suggested throughout this book is that "we the people" are increasingly using innovations to reclaim some of that authority as political systems fail us. We are using new technological capabilities to express our desires and to push back against authorities who have sometimes lost touch with common sense or who have too long taken the consent of the governed for granted. Moreover, I believe that innovation represents a reasonable way to reclaim some of our consent, especially compared with more radical steps, which would be disruptive, destructive, and unwise. We might even think of this process of innovation-as-dissent as the equivalent of a slow-motion revolt: Change is coming not in one fell swoop but instead through incremental acts of technologically enabled creativity and resistance.

In this sense, as I argue next, innovation increasingly gives us greater voice and exit options to push back against authority. We may sometimes go even further and use technologies to live lives of our own choosing.

Innovation as "Voice" and "Exit"

In his 1970 book, *Exit, Voice, and Loyalty*, the economist Albert O. Hirschman discussed the interplay between voice and exit—for businesses, organizations, and even governments. He began his book with the following observation:

> Under any economic, social, or political system, individuals, business firms, and organizations in general are subject to lapses from efficient, rational, law-abiding, virtuous, or otherwise functional behavior. No matter how well a society's basic institutions are devised, failures of some actors to live up to the behavior which is expected of them are bound to occur, if only for all kinds of accidental reasons. Each society learns to live with a certain amount of such dysfunctional or mis-behavior; but lest the misbehavior feed on itself and lead to general decay, society must be able to marshal from within itself forces which will make as many of the faltering actors as possible revert to the behavior required for its proper functioning.[51]

Hirschman went on to discuss how "repairable lapses" might be best addressed by organizations (firms or governments) and to what degree the loyalty of individuals (as customers or as citizens) might affect that balance. In this regard, Hirschman argued that there is an important interplay between voice and exit that is often underappreciated. "The chances for voice to function effectively as a recuperation mechanism," he argued, "are appreciably strengthened if voice is backed up by the *threat of exit*, whether it is made openly or whether the possibility of exit is merely well understood to be an element in the situation by all concerned."[52]

Creative acts and the technological innovations they give rise to can represent both the voice and the exit options that Hirschman described. When individuals—either on their own or as part of a collective effort—set out to create new and better ways of achieving various ends, they are challenging some sort of status quo. They will quite often first use their voice to ask others, "Why can't things work better?" or "Why can't we solve this problem?" When they discover that that status quo makes it difficult to answer those questions, they sometimes use the threat of exit to challenge existing norms—social, business, or governmental.

Again, as Chapter 3 discussed, innovative acts might best be considered a type of "exit" that falls short of the more radical kind. In this sense, ongoing innovative activities operate as a sort of relief valve, dissipating pressures that have built up in a regulatory system. To use another analogy, we might even liken evasive activities to a sort of "controlled burn" that helps ensure the innovation ecosystem can prosper over the long term.[53] Just as controlled burning represents a form of hazard reduction to prevent catastrophic forest fires, tolerating and learning from evasive entrepreneurialism can have a beneficial effect on regulatory systems and the private social systems they govern.

This book has documented various examples of technologically enabled *marginal* evasiveness by entrepreneurs that facilitate that end. In almost all the cases, complete "exit" from a polity or regulatory system was never in the cards for most innovators. Instead, evasive entrepreneurs operated at the margin to challenge existing norms that they found outmoded, inefficient, or illogical. Again, the process is more akin to incremental protest or slow-motion rebellion. Here is how my former colleague Eli Dourado described it in a 2011 essay:

> The productive people do not go on strike when they are over-controlled. Instead, they innovate around the points of control. They go on strike *at the margin*. And it doesn't take a big, dramatic exit. A little bit cumulatively over decades is sufficient to both be noticeable in the data and to reduce the amount of control that can be exercised.[54]

Let's return to some of our case studies from Chapter 2 and consider how evasive entrepreneurs used voice or exit to go on strike at the margin and how it affected the status quo. The sharing economy example is worth revisiting in this regard. We can see Hirschman's insights at work when innovators use the threat of exit to leave a jurisdiction or evade certain preexisting regulations. Think back to the late 2000s, when sharing economy platforms Uber and Airbnb were being launched. Though sometimes vilified for the way they pushed the envelope and used evasive strategies to gain leverage in regulatory negotiations, consider the alternatives these innovators faced had they not done so.

For many decades, economists and other reform-minded policy advocates had made the case for liberalizing local transportation and hospitality sector regulations that were hopelessly inefficient and that hurt competition and consumer choice. Yet reformers made little progress despite the weight of their evidence. Markets remained largely closed to competition, and incumbents were protected against new entrants and innovations. Consumers bore the cost of these foolish policies in the forms of fewer choices, higher prices, and poorer-quality service. Again, this dismal state of affairs represented the norm for many decades.

Then Uber and Airbnb came along and changed the entire conversation in just a few short years. Had Uber and Airbnb not taken the paths they did, would we even be talking about this sector and all these innovations right now? People didn't realize what they were missing until entrepreneurs went out and created it. Today, no one except for old taxi and hotel companies would advocate an outright ban on such competitors, even if those companies started out life by playing by a different set of rules. By pushing back against the permission society and blazing their own trails through acts of evasive entrepreneurialism, these firms gained the leverage to meet regulators on better terms and offer citizens important new options they did not have previously.[55]

This interplay between measures and countermeasures represents a beneficial tension in that the threat of exit or evasion strengthens the voice of those not previously heard. When innovators and consumers can gain leverage through such threats, it puts pressure on governments to be more responsive. This tactic can provide a useful check on the worst tendencies of overzealous regulators and lawmakers. We might already be seeing that positive pressure reflected in the more flexible ways some cities are dealing with rental scooters and bike sharing. Barriers to such micromobility competition still exist, and some governments are cracking down on them, but many others seem to be allowing innovation to take root even before clear rules have been worked out.

A Dance That Never Ends

There is no endpoint in this tit-for-tat process of measures and countermeasures, and it is impossible to forecast who will win in any particular case. Nor can we even define what winning means in many cases. My colleague Andrea O' Sullivan explains:

> Technology and the state are caught in a constant cat-and-mouse game of surveillance and evasion. As soon as one side gets wise to the other's tricks, they modify tactics to again outpace their rival. Rinse and repeat until either everyone is in jail or law enforcement just gives up. Then the next technological phase change emerges and the race starts all over again.[56]

One thing is certain, however: these technological races will multiply and accelerate. Debora L. Spar opened her important 2001 book, *Ruling the Waves: Cycles of Discovery, Chaos, and Wealth from the Compass to the Internet*, by noting that "there is a certain give-and-take along the technological frontier, a dance of regulation that moves power back and forth between firms and government, between pioneers and bureaucrats."[57] Her impressive histories of various technology sectors showed how this dance played out in the form of four waves that she labeled *Innovation*, *Commercialization*, *Creative Anarchy*, and *Rules*. Importantly, however, Spar presciently concluded her tome with the prediction that these cycles would unfold more rapidly as the pace of technological change intensified. Indeed, that is precisely what is happening today as the pacing problem accelerates and evasive entrepreneurs are empowered by new technologies of resistance. In essence, there are now many dances going on at the same time and more parties participating in those dances than ever before.

Globalization and innovation arbitrage will continue to be major contributing factors to these dances, as Chapter 2 documented. With emerging technologies evolving rapidly and governments across the globe competing to attract new investment and expand job opportunities, arbitrage opportunities will likely continue to multiply.

The ramifications of such developments for government are not always clear cut, however.[58] Under one scenario, when innovators shop around or vote with their feet more actively, it creates positive pressure on governments to behave in a more sensible, streamlined, decentralized fashion. The result, as Alfred Aman has suggested, is that "the processes of globalization can weaken the state in various ways, not the least of which is that they make it relatively easy for some industries to move production around the globe."[59] Samuel Hammond of the Niskanen Center has documented how digital technologies helped usher in restructuring in both the corporate and public sectors in the 1990s. "The next wave of e-government reforms promises to be a tsunami, opening the highest levels of the state apparatus to technological disruption," thanks to innovations like blockchain, smartphones, and more sophisticated application programming interfaces married to AI or machine-learning systems.[60] These innovations could lead to more decentralized and competitive governance systems and approaches, Hammond believes.[61]

Under another scenario, accelerated arbitraging could also lead to the sort of *Empire Strikes Back* moments discussed earlier, with policymakers advocating even more repressive regulatory approaches to constrain evasive activities. In addition, pressures could mount for federal or international harmonization that largely forecloses arbitrage opportunities at lower levels of government.

My hope is that we can find a sensible balance and that these trends will have a corrective or refreshing effect on government policies. As Hirschman argued, "exit has an essential role to play in restoring quality performance of government, just as in any organization."[62] To reiterate, expanded "exit" opportunities and efforts do not mean government will magically shrink or go away. That is fantasy thinking that we can dispense with. But just as innovation is fueled by incentives, acts of innovation that are evasive in character can incentivize governments to behave differently and devise smarter policies. Innovation-as-exit can be viewed as a threat, but my hope is that it will be instead viewed as an opportunity to pursue positive dialogue and reforms that will make governments more accountable and adaptive.

Indeed, if competition and trial-and-error experimentation are good for science, for markets, and for just about everything else, then why are they not also good for government programs and functions? "I think almost every economic system can benefit from experimentation with new forms of government and new types of government services," argues Paul Romer, who recently won a Nobel Prize in economics.[63] Romer is exactly right. Alas, the status quo all too often prevails. If innovation and evasive entrepreneurialism can help shake things up a bit, we should be willing to see how that process plays out.

6

HUMANISM, ETHICS, AND RESPONSIBLE INNOVATION

Thus far, this book has mounted a defense of technological innovation and argued that even acts of evasive entrepreneurialism have great value to society. But there are downsides to them, too. The next two chapters respond to some common objections about the ethical or human dimensions of technological innovation, specifically the allegation that innovation somehow undermines our humanity.

In response, I will show that technological innovation is fundamentally about improving our humanity by bettering our lives as well as the lives of those around us and even those far away from us. Properly understood, technology and humanism are complements, not opposites. I will also explore the tension between permissionless and responsible innovation and argue that not only are these concepts compatible, but also they are already being balanced through a variety of governance practices sometimes referred to as soft law, as defined in the Introduction and discussed in Chapter 2.

Adaptation in the Face of Adversity

As we have seen, the future is unfolding more rapidly than law's ability to keep up with it. Some techno-optimists vociferously cheer these developments and predict a better world will follow. By contrast, a great many techno-pessimists loathe them and live in fear of the supposed dystopian hellscape to come.

I obviously lean strongly in the techno-optimists' direction in most cases. My perspective is based not on blind faith in technology but rather on a factual evaluation of what has, again and again, lifted humanity to new heights throughout history. By creating new tools to solve basic problems and fulfill important needs, technological innovation has improved human well-being.[1]

The pessimists have some legitimate concerns about the potential dangers associated with new technologies, however, especially as traditional governance systems break down and the potential for more widespread legal evasion grows. Although some of their concerns are understandable, others, and their proposed solutions, often leave much to be desired.

For example, many critics today decry the frictionless nature of modern innovations and suggest that society should hit the pause button on technological developments, or at least find reasonable ways to slow things down a bit.[2] It is fine and well to add friction *voluntarily* to one's personal routine in an attempt to achieve a better balance with modern technologies.[3] It is an entirely different matter, however, to suggest that friction should be forced upon us through coercive mechanisms or even regulatory nudges that restrict our options and opportunities. I have already made it clear why it would be a serious mistake to throw a wrench in the gears of progress in that fashion. Technological innovation is the fundamental driver of human betterment over time. Stopping or even slowing the rate of technological change is a call for stasis, and stasis will be our ruin as a species. It would have a profoundly negative effect on economic growth, living standards, our health and welfare, and our personal autonomy.[4]

Even when the potential for some harm exists and the case for adding friction in the form of some sort of regulatory intervention may be stronger, it does not mean the proposed remedies will work or be cost-effective. It is always easy for technological critics and concerned policymakers to insist that something must be done when a new technology is speeding ahead. It is quite another thing for critics to devise workable policies that won't result in enormous costs for society and discourage the development of innovations that could significantly improve our quality of living. Virtually every major regulatory agency is already grappling with this problem today as these technological capabilities expand and acts of evasive entrepreneurialism increase.

Still, the critics have a fair point: The pace of innovation feels overwhelming at times, and it truly does come into conflict with ideas and institutions that have great importance to many people. How, then, can all those concerns be addressed to ensure that humanist values are preserved in an age of rapid technological change?

The short and honest answer is that not everything can be perfectly addressed to make those critics happy. Some ideas and institutions will need to adapt. Yet values have *always* evolved throughout the course of civilization.[5] That does not mean that innovation should (or will) sweep away all that is old. The more sophisticated answer to the above question is that society typically finds a way to find balance, adapt, and muddle through.[6]

We humans are a remarkably resilient species, and we regularly find creative ways to deal with major changes through constant trial-and-error experimentation and the learning that results from it. In that process, we find a new baseline or equilibrium and incorporate new ideas, institutions, and values into our lives. We will continue to do so, but it will not always be according to the sort of script that many critics desire.

Humanism, Technology, and the Specter of Determinism

Because many of the calls for responsible innovation flow from humanist critiques of technological innovation, it is important to first address what some mean by "humanist." As Chapter 1 noted, there exists no shortage of critics who label themselves humanists and who decry the supposedly deleterious effects technological change has on individuals, institutions, or culture. Humanist critiques of innovation can be different from each other, but a core attribute of most of them is the notion that technology and technological change are somehow at odds with humanity and human flourishing. Accordingly, critics regularly use terms like dehumanizing or "re-engineering humanity" when discussing their fears about new technologies.[7] I have already labored to prove just how off base those critiques are, but it is worth diving a little deeper to understand why technological innovation and human flourishing are complements, not enemies.[8] First, though, we must address the accusation that defending innovation represents little more than an acceptance of a technological future that is devoid of any concern about other important values.

Across the field of Science and Technology Studies (STS), scholars have long decried what is known as "technological determinism." Generally speaking, technological determinism is defined as the belief that "technological developments take place outside society, independently of social, economic, and political forces" and that "technological change causes or determines social change."[9] The opposite of technological determinism is referred to as "social determinism" or "social constructivism," which "presumes that social and cultural forces determine technical change."[10]

In STS discussions, to be labeled a technological determinist these days is akin to being affixed with a scarlet letter of shame. It implies that you are a naïve technology booster who sees no role for politics, society, or average people in shaping their own destinies. Technological determinism, as defined by its many critics, represents the height of anti-humanist thinking. Those critics have also come up with many creative labels to describe the same notion, including "technologism,"[11] "techno-fundamentalism,"[12] "technological solutionism,"[13] and even "techno-chauvinism."[14] Regardless of the monikers the critics choose to decry technological determinist thinking, they are unified in thinking that "people-based solutions" represent the morally superior approach to ensuring that future populations will live in a "people-centered economy."[15]

Those critics are creating a false dichotomy. When tech critics play the humanist card, they seem to imagine that they somehow have nobler intentions and a deeper concern for the plight of people than others do. Meanwhile they attack those who dare suggest that technological change has been a core driver of human betterment, even though it is an unambiguous fact. Consider the way technological determinism is typically described in STS literature. Sally Wyatt articulates the common conception of deterministic thinking as follows:

> One of the problems with technological determinism is that it leaves no space for human choice or intervention

and, moreover, absolves us from responsibility for the technologies we make and use. If technologies are developed outside of social interests, then workers, citizens, and others have very few options about the use and effects of these technologies.[16]

Framed in that fashion, it is completely understandable why critics would lambaste anyone adhering to such a worldview. In reality, however, few people hold such an extreme view about technology being the only important force shaping the course of history or human affairs.

What is particularly ironic is that some of the most rigid technological determinists are technology critics themselves. "A primary characteristic of the antitechnologists," Samuel Florman once argued, "is the way in which they refer to 'technology' as a thing, or at least a force, as if it had an existence of its own" and which "has escaped from human control and is spoiling our lives."[17] For example, some of the most notable tech critics of the past half century were French philosopher Jacques Ellul, American historian Lewis Mumford, and American cultural critic Neil Postman. Their books painted a dismal portrait of a future in which humans were subjugated to the evils of "technique" (Ellul),[18] "technics" (Mumford),[19] or "technopoly" (Postman).[20] The narrative of their works read like dystopian science fiction books. Essentially, there was no escaping the iron grip that technology had on us. Postman claimed, for example, that technology was destined to destroy "the vital sources of our humanity" and lead to "a culture without a moral foundation" by undermining "certain mental processes and social relations that make human life worth living."[21]

When dour tech critics like these preach the gospel of technological gloom-and-doom, they usually get a free pass from their fellow tech critics despite the clear deterministic overtones. Apparently it is acceptable to use deterministic reasoning when your intentions are "pro-human" and your preferences are in line with other innovation critics. If, however, one dares employ any sort of deterministic arguments when speaking *optimistically* about the future, that person is decried as uncaring and anti-human.

Generally speaking, we can dismiss extreme deterministic reasoning—regardless of whether it's tech optimists or pessimists making such claims—for a rather simple reason: technologies fail all the time. "If promising technologies can suffer fatal blows from unexpected circumstances," Florman correctly argued, then "[t]his means that we are still—however precariously—in control of our own destiny."[22]

Technologies fail for many reasons, but societal demands and citizen pushback are two underappreciated explanations for why so many technologies flounder or are rejected. For example, in 2013, Google launched Google Glass, a pair of augmented reality "smart glasses" that would let users access information about their surroundings via a pop-up interactive display. Within two years, however, Google had canceled the project for consumer use and instead moved to offer a version of Glass only for commercial enterprises to use for specific workplace tasks. Perhaps Google Glass failed because of its hefty $1,500 price tag, or maybe there was not much consumer need for such a product yet. An equally compelling explanation for the failure of Google Glass was the "creepiness" factor associated with it. Privacy advocates decried the device and critics used the derogatory term "Glassholes" when referring to Glass users.[23] This product was an example of what Nobel Prize–winning economist Alvin E. Roth once referred to as "repugnance as a constraint on markets."[24] The intensity of the public backlash forced Google and other augmented reality companies to reconsider the wisdom of wearable smart glasses. "If the stigma surrounding Google Glass (or, perhaps more specifically, 'Glassholes') has taught us anything," argued *Wired* journalist Issie Lapowsky, "it's that no matter how revolutionary technology may be, ultimately its success or failure ride on public perception. Many promising technological developments have died because they were ahead of their times."[25] A similar sort of public repugnance about new facial recognition technologies appears to be growing and could limit the diffusion of that technology.[26]

This example shows why deterministic thinking is too simplistic—people push back against technology all the time, and tools are constantly being reformed to better suit our collective desires and demands. We need a more balanced perspective in these debates. For lack of a better term, we might think of the middle-ground position as "soft determinism." That is, one can believe that technology plays an important role in influencing history—and that innovation oftentimes moves faster than law's ability to keep pace with it—while also believing that society, governments, and each and every human being can and will play a major role in shaping technology's nature and evolution. Others have defined soft determinism as the idea that "technological change drives social change but at the same time responds discriminatingly to social pressures."[27]

Although soft determinism represents a more reasonable position in these debates—and one that also offers a more realistic explanation of how technological governance works in practice—a great many scholars and policy advocates continue to heighten their approval of humanist labels and rhetoric. But what exactly is a humanist critic?

Most self-declared humanist scholars would probably agree with philosopher L. M. Sacasas that "[h]umanism is a rather vague and contested term with a convoluted history."[28] To some extent, humanist critiques of technology are simply meant to remind us that all people are important, as is the case when some claim the humanist position represents "a philosophical claim about the centrality of humankind to the universe."[29] Again, who could be against such an assertion, or the repeated claim made by other self-anointed humanists who insist technological change has many tradeoffs and downsides? In a 2015 essay, Andrew McAfee of the MIT Sloan School of Management noted that such observations are uncontroversial and widely agreed upon.[30] The problem, he correctly noted, is that such banalities should not be used to end any inquiry into the benefits of technological change. Unfortunately, that is exactly what often happens in the field of science and technology scholarship and policymaking today. McAfee describes this attitude as follows:

> The third sense of "humanist" is by far the most problematic. It's close to: "Because I am for the people I should be free from having to support my contentions with anything more than rhetoric." Or, more simply: "You can trust what I say, because I am on the side of people instead of the cold, hard machines." Well, no. We should evaluate what you say based on the quality and quantity of evidence you've marshaled, and on the rigour with which you have analysed and presented it. If this sounds like an argument in favour of the scientific method, that's because it's exactly what it is.[31]

As McAfee suggests, critics who insist that technological innovation is anti-human or dehumanizing and use such rhetorical ploys to reject a particular innovation bear some burden of proof of the alleged harms. They must be willing to acknowledge that there are tradeoffs associated not only with new technologies, but also with the remedies they propose to any alleged downsides.

As I discussed in my previous book, those who advocate slowing or stopping technological advances need to demonstrate that the harms they allege are highly probable, tangible, immediate, irreversible, catastrophic, or directly threatening to life and limb in some fashion.[32] In recent years, risk analysis tools have improved and cost-benefit analysis has become formalized within the regulatory policymaking process. These tools and methods can be used by those advocating preemptive, prohibitive controls on new tech.[33] Oftentimes, as will be noted later, the critics do not bother spelling out what sort of remedies they think are appropriate. They feel it is enough to decry the supposed downsides associated with technology, suggest that "something must be done," and then presumably expect someone else (usually government actors) to take up that cause. That is where their analysis all too often ends. Little effort is put into exploring the full range of tradeoffs associated with the various (but unspecified) innovation-limiting actions they argue are needed.

At worst, tech critics sometimes rest their case for limiting innovation on nostalgic arguments about some proverbial good old days—all the while deftly avoiding telling us precisely when those days were. The problem with all the punditry in what Richard Posner once aptly labeled "the declinist genre" is that it is flatly at odds with the actual historical record regarding the state of human affairs in the past.[34] Even a cursory review of history offers voluminous, unambiguous proof that the old days were, in reality, eras of abject misery. Widespread poverty, mass hunger, poor hygiene, short lifespans, and so on were the norm. What lifted humanity up and improved our lot as a species is that we learned how to apply knowledge to tasks in a better way through incessant trial-and-error experimentation.[35] In other words, humanity flourished by *innovating*, and the results of our innovative activities were called *technologies*.

Technology is not some mystical force that appeared out of thin air. Nor is it an autonomous entity with a will of its own. *All technology is the product of human design and action.*[36] The most straightforward definition of "technology" is simply the application of knowledge to a task, and as Benjamin Franklin once noted, man is a tool-making animal by his nature. "[T]he elementary pleasure of solving technical problems and successfully completing constructive projects," Samuel Florman once correctly observed, is "as old as the human race."[37]

Thus there are few things more humanist than crafting tools to solve important problems and to better our lives and the lives of our loved ones and others.[38] One can simultaneously believe in "the centrality of humankind to the universe" as well as the notion that technological innovation is central to humankind's ability to improve the little corner of the universe that we occupy.

How Technology Expands the Horizons of Our Humanity

Technology helps us better understand and address the needs of strangers at a distance. In his 1759 *Theory of Moral Sentiments*, the Scottish moral philosopher and economist Adam Smith observed the following:

How selfish soever man may be supposed, there are evidently some principles in his nature, which interest him in the fortunes of others, and render their happiness necessary to him, though he derives nothing from it, except the pleasure of seeing it. Of this kind is pity or compassion, the emotion we feel for the misery of others, when we either see it, or are made to conceive it in a very lively manner. That we often derive sorrow from the sorrows of others, is a matter of fact too obvious to require any instances to prove it; for this sentiment, like all the other original passions of human nature, is by no means confined to the virtuous or the humane, though they perhaps may feel it with the most exquisite sensibility. The greatest ruffian, the most hardened violator of the laws of society, is not altogether without it.[39]

Smith believed that humans were both self-regarding and other-regarding and that we had an innate moral sensibility and sympathy for others, or what he called a "fellow-feeling." Thanks to this natural sensibility, we would first look to take care of ourselves and those closest to us, but we would then look to help others the best we could.[40]

During Smith's time, however, that "fellow-feeling" for the plight of others was limited by social, economic, and technical realities. Most people were confined to the family farm or working in a small shop or later in a factory in town. They were also unable to travel far beyond their immediate communities. Communication technologies did not yet give them the ability to learn much about the world beyond their own communities, except perhaps through newspaper accounts or secondhand information that trickled in weeks or months after developments occurred elsewhere.

Flash forward two centuries and consider how technology, in the words of American historian Thomas L. Haskell, "change[d] the moral universe in which we live."[41] In a two-part 1985 essay on the "Origins of the Humanitarian Sensibility," Haskell observed how "our feeling of responsibility for the stranger's plight, though nowhere near strong enough to move us to action, is probably stronger today than it would have been before the airplane."[42] The growth of ubiquitous, affordable transportation and other technological capabilities—most notably widespread, instantaneous communication and information transmission—has expanded our moral universe. Haskell argued the following:

> Technological innovation can perform this startling feat, because it supplies us with new ways of acting at a distance and new ways of influencing future events and thereby imposes on us new occasions for the attribution of responsibility and guilt. In short, new techniques, or ways of intervening in the course of events, can change the conventional limits within which we feel responsible enough to act.[43]

By constantly expanding the horizons of our moral universe in this fashion, technology expands our humanitarian sensibility. It enables us to be more worldly, cosmopolitan, and compassionate. "The Humanist Manifesto," originally published by the American Humanist Association in 1933 and most recently updated in 2003, asserts that humanists "ground values in human welfare shaped by human circumstances, interests, and concerns and extended to the global ecosystem and beyond. We are committed to treating each person as having inherent worth and dignity, and to making informed choices in a context of freedom consonant with responsibility."[44]

This is a noble vision of life and living, but it should also be clear why innovation is central to that humanist narrative. Innovation is central to human betterment not simply because it betters *us*, but because it allows us to better the lot of our fellow humans. Innovation expands our responsibility for each other and allows us to better act upon our "fellow-feeling." "Progress consists of deploying knowledge to allow all of humankind to flourish in the same way that each of us seeks to flourish," notes Steven Pinker.[45] Technology helps us achieve this goal and enhances our humanity by helping us understand and address the needs of our fellow humans across the globe, many of whom we will never meet. Again, what could be more humanist than that?

Making Permissionless Innovation and Responsible Innovation Compatible

Although humanist critiques of technology often go much too far, innovation's defenders should take seriously calls by critics to incorporate other values or rights into the process of technological development and governance. Concerns about the safety, security, and privacy-related implications of many emerging technologies are particularly notable in this regard because those issues pervade almost every emerging technology sector today.

Those concerns have led to a growing intellectual movement known as "responsible research and innovation" (RRI). Although this movement is more widespread in Europe, it is growing in the United States, but sometimes under the auspices of "technology ethics" or other labels.[46] In the United States, the term "upstream governance" is often used to refer to largely the same thing.[47] A great deal of work by STS scholars today revolves around these themes of "responsible innovation," "ethical innovation," and "upstream governance."[48] Definitions are still evolving, but a 2011 article by René von Schomberg, a leader in the RRI movement and the Director General for Research at the European Commission, defined RRI as follows:

> A transparent, interactive process by which societal actors and innovators become mutually responsive to each other with a view to the (ethical) acceptability, sustainability and societal desirability of the innovation process and its marketable products (in order to allow a proper embedding of scientific and technological advances in our society).[49]

Other scholars define RRI more simply, saying that it comes down to "taking care of the future through collective stewardship of science and innovation in the present."[50] Practically speaking, this definition means anticipating the potential adverse consequences associated with technological change and seeking to somehow mitigate them through some form of upstream governance.

In a sense, RRI is just an extension of corporate social responsibility (CSR), a widely discussed but quite amorphous concept in the United States and abroad. It is not really clear what CSR means in many contexts, or that it even works that well in practice.[51] Regardless, CSR has become a major part of modern business practices and decisionmaking. RRI builds on CSR, but RRI is more squarely focused on addressing the potential risks associated with specific technologies or technological processes. In 1970, the Nobel Prize–winning economist Milton Friedman observed that discussions about CSR "are notable for their analytical looseness and lack of rigor."[52] His statement is still somewhat true for CSR today, and it is especially true for RRI. Both concepts remain open to differing interpretations and incorporate many distinct values that vary by context. At root, however, what RRI and CSR have in common is the belief that, whatever those responsible values are, they should be baked in early during product decisionmaking and design.

At first blush, it may seem as if permissionless and responsible innovation are fundamentally at odds. To the contrary, RRI can very much be part of a policy regime that adopts permissionless innovation as its general tech policy default. These concepts can coexist so long as policymakers and RRI advocates are willing to think more broadly about what the term "governance" means as applied to technological processes. "Governance" can mean more than just formal regulation by legislatures, administrative agencies, or other public bodies. Governance can also describe a much broader universe of norms and rules that are established and enforced by a wide variety of people (or groups of people) in a wide variety of ways.

When we consider questions of *technological* governance—and specifically the notion of anticipatory governance, which is a prominent feature of RRI discussions—it helps to specify whether we are speaking of governance in a broad or narrow sense. Whether consciously or not, RRI scholars and advocates often fail to make clear precisely what type of governance they desire. This distinction is important because if anticipatory governance involves the formal application of the precautionary principle by force of law, it will be a deal-breaker for many innovation advocates. Banning innovative acts on the basis of fears about hypothetical worst-case scenarios means that many best-case scenarios can never come about. It forecloses trial-and-error experimentation that could bring about life-enriching services and applications. For example, many tech critics suggest that robotic technologies should be preemptively regulated based on a host of worst-case *Terminator*-esque scenarios about killer machines or AI run amok. Those far-fetched fantasies make for great sci-fi stories, but meanwhile back in the real world, robotic exoskeletons are helping people with spinal injuries walk again, and inventors are working to create autonomous vehicle technologies that could save countless lives in the future. That is why worst-case thinking should not guide policy and why innovation should be innocent until proven guilty.

Precaution can be pursued through less restrictive approaches, however. Some government agencies allow innovations to be released into the wild but in accordance with established safety standards and with a recall regime for defective or unsafe products. For example, this is the way the National Highway Traffic

Safety Administration addresses motor vehicle safety and the Consumer Product Safety Commission deals with unsafe devices. This represents a softer form of precaution relative to harder constructions of the precautionary principle, such as outright bans.[53] Even less restrictive but still precautionary in orientation would be a mandatory labeling law or a government-led risk reduction educational campaign. The Food and Drug Administration uses that approach in many instances. As noted earlier and discussed at greater length later, soft-law governance approaches are also used in many sectors today. Soft law includes various tools and methods—multistakeholder processes, agency guidance documents, collaborative best practices, industry standards and self-regulation, and so on—that establish expectations about technological development or use but that lack the same level of enforcement that accompanies hard-law enactments.

In other words, there exists a broad spectrum of governance options for new technologies, and it is important to specify what sort of approach we are talking about when debating these issues. We will return to soft-law mechanisms in the next chapter and explain how they can help us find a sensible balance in tech governance discussions.

Responsible Innovation without the Precautionary Principle

Exactly how much formal upstream governance of a *regulatory* nature does the responsible innovation movement recommend? It is often not very clear. Many responsible innovation advocates seem sympathetic to policies based on the precautionary principle and highly skeptical of the wisdom of permissionless innovation as a policy default. Yet most of them never spell out the exact relationship between RRI and the precautionary principle as a matter of public policy.

Some advocates of responsible innovation argue that the focus "is more on mitigating wider societal long-term risks and so [RRI] favors incremental rather than radical innovation."[54] That focus suggests a closer connection between RRI and a formal application of the precautionary principle for emerging technologies. A 2015 Brookings Institution white paper by Walter D. Valdivia and David H. Guston provides a more concrete answer to this question. Valdivia and Guston insist that responsible innovation "is not a doctrine of regulation and much less an instantiation of the precautionary principle; the actions it recommends do not seek to slow down innovation because they do not constrain the set of options for researchers and businesses, they expand it."[55]

Unfortunately, this demarcation between the general notion of responsible innovation and the formal application of the precautionary principle is not nearly so well defined in most RRI literature. On the rare occasions when RRI proponents *do* define the line between them, the proponents often introduce other terms that are equally amorphous. Even Valdivia and Guston fall prey to this tendency when they go on to suggest that responsible innovation "considers innovation inherent to democratic life and recognizes the role of innovation in the social order and prosperity," but that RRI advocates desire "a governance of innovation where that choice is more consonant with democratic principles."[56] The problem is that this notion simply shifts the definitional challenge away from defining "responsible innovation" and toward a debate about how we define "democratic life" and "democratic principles" in any given context.

The RRI literature is rife with ambiguous terms such as those and many others, such as "the public interest," and yet these advocates consistently lack precision regarding what they mean and how their claims can be translated into concrete governance principles or policies.[57] For example, does making innovation more "consonant with democratic principles" mean that each new technology is somehow subjected to a formal vote before it gets released? If so, it would be hard to imagine many important innovations ever seeing the light of day.

Even if they cannot be nailed down on the precise applicability of the precautionary principle within technology policy debates, most RRI advocates would reject permissionless innovation as a suitable default position. Some of them even imply that permissionless innovation is synonymous with anarchy or a complete disregard for human rights. But that stance is ludicrous. For my book about permissionless innovation, I surveyed almost countless essays and articles that cite the phrase. Not once did I see any advocate of permissionless innovation going to such extremes as these critics suggest. Perhaps even more surprising is that most advocates of permissionless innovation rarely propose abolishing laws or agencies that currently oversee existing technologies or sectors.

What permissionless innovation advocates generally are advancing is the notion that new ideas deserve a fair shake or that *entrepreneurs and innovations should generally be considered innocent until proven guilty*. Permissionless innovation means giving innovators a bit more breathing room and avoiding a knee-jerk rush to regulate the new and the different. It means innovators should have a green light to experiment with those new and different ideas unless we can agree that a compelling reason exists to disallow trial and error as the basis of innovation policy.

The precautionary principle, by contrast, recommends keeping the light red until innovators can prove that new products and services are perfectly safe, however that is defined. But there are many points along the spectrum between these two policy postures and even many values shared by advocates of both perspectives. The difference will often come down to the processes we use to address the areas of difference.

A Willingness to Compromise

Properly understood, responsible innovation can be compatible with the permissionless governance vision and policy regimes that have the freedom to innovate as their operational default. To achieve that synthesis, however, those on both sides must agree to some compromises. Advocates from different perspectives need to be open to learning from each other and willing to take the other's concerns seriously. Flexibility is essential.

To begin, RRI advocates need to appreciate how they can accomplish a great deal of good even in the absence of formal regulatory action. Their first instinct should not be to decry permissionless innovation advocates as a bunch of uncaring anarchists. If that is their starting point in conversations about emerging tech, RRI advocates will be missing opportunities to work with diverse parties and instill wise principles into various technological development processes. Such a move would be particularly misguided during a time when the pacing problem has become an undeniable reality and has made traditional hard-law efforts more difficult.

If they hope to get some (or perhaps any) of the values and procedures that they care about incorporated into technological development processes, RRI advocates will need to be open to the idea that perhaps the only way to do so will be through less formal procedures, precisely because law will likely lag so far behind marketplace developments. They should also appreciate the limitations of traditional regulatory approaches and the deleterious effects those regimes have sometimes had on innovation and competition. Again, that will require compromise.

Many such scholars now speak of the need for new forms of technological governance "that move beyond traditional command-and-control policymaking and enforcement to improve the effectiveness and legitimacy of regulation."[58] "A good governance approach," notes Schomberg, "might be one which allows flexibility in responding to new developments."[59] He writes:

> The power of governments is arguably limited by their dependence on the insights and cooperation of societal actors when it comes to the governance of new technologies: the development of a code of conduct, then, is one of their few options for intervening in a timely and responsible manner.[60]

That sort of thinking will be required among RRI advocates who hope to find common ground with permissionless innovation advocates.

Permissionless innovation advocates will need to be open to new ideas and perspectives, too. If the first instinct among them is to dismiss the entire RRI movement as little more than repackaged Luddism, hell-bent on derailing all the great inventions of the future, then permissionless innovation advocates are foolishly forgoing the chance to work with a diverse group of well-intentioned scholars and stakeholders who could ensure that new products and services gain more widespread acceptance and public trust. More practically, those who support permissionless innovation would be wise to accept that, although technological innovation is oftentimes outpacing the ability of government to keep up, well-established regulatory regimes or agencies are not necessarily going away any time soon. To repeat, few technocratic laws or regulatory bodies have been liberalized or eliminated in recent memory. It is unlikely that trend will reverse any time soon.

Thus, taking responsible innovation priorities seriously—and finding flexible ways to instill them in the development process— offers permissionless innovation advocates a chance to forge a rough peace with policymakers and issue advocates who often just want to have a small say in how technological processes are unfolding. But, if regulators seek to have a *big* say in such matters—namely, in the form of heavy-handed, preemptive restrictions—then major policy fights will no doubt ensue.

But responsible innovation advocates must certainly understand that the era of technocratic, overly bureaucratic, top-down, command-and-control regulation is being challenged by new realities.[61] Philip Weiser notes that "[t]he traditional model of regulation is coming under strain in the face of increasing globalization and technological change," and, therefore, governments must think and act differently than they did in the past.[62] "The new information environment," argues Taylor Owen, "may require states to adopt some characteristics of start-ups."[63] Similarly, Juma hoped to see "entrepreneurialism exercised in the public arena."[64]

The next chapter explores how entrepreneurial governance approaches are emerging today in the form of soft-law mechanisms. These mechanisms offer the greatest hope for compromise and sensible governance of emerging technology.

7

SOFT LAW AND THE FUTURE OF TECHNOLOGICAL GOVERNANCE

The previous chapter concluded with a call for compromise between responsible and permissionless approaches to innovation. This chapter shows how, as a practical matter, such compromises are already being negotiated in many policy deliberations about emerging technology governance through what has come to be known as soft law.

Soft law represents a messy amalgam of many different governance approaches that will often leave all sides somewhat dissatisfied because they will not get everything they want. That dissatisfaction, however, might be the best thing going for soft law. Much as Winston Churchill once famously said that democracy represented "the worst form of Government except for all those other forms that have been tried from time to time," it may be the case that soft law represents the worst form of technological governance, except for all those tried before.

This chapter explores what is meant by soft law. It explains how soft law is already becoming the dominant approach for modern technological governance (at least in the United States), and it shows how soft law is supplementing existing legal and regulatory hard-law remedies. Also explored will be the role that various other expert organizations play in facilitating technological governance today, as well as the importance of stepped-up risk education efforts in addressing various concerns. Finally, I discuss the shortcomings of soft law and more challenging issues regarding technological risks that might require more serious regulatory oversight.

Soft Law: The Basics

Soft-law mechanisms have already been mentioned many times throughout this book. Soft law includes a wide variety of informal, collaborative, and constantly evolving governance mechanisms that differ from hard law in that they lack the same degree of enforceability.[1] These soft-law systems and processes are multiplying at every level of government today: federal, state, local, and even global.

The easiest way to define soft law is to explain what it isn't: Soft law is not hard law. Soft law builds upon and operates in the shadow of hard law. But soft law lacks the same degree of formality that hard law possesses. Scholars who study such governance mechanisms note that soft law is used "as a shorthand term to cover a variety of nonbinding norms and techniques for implementing them."[2] Although some consider this informality and nonbinding nature to be a weakness of soft law, it also serves as a strength. Compared with hard law, soft law can be more rapidly and flexibly adapted to suit new circumstances and address complex technological governance challenges.[3]

Chapter 4 identified the many reasons that evasive entrepreneurialism is on the rise today. To reiterate, those problems include the accumulation of laws and regulations that are increasingly out of touch with common sense, a chronic inability of government institutions to reform broken governance systems or adapt regulatory policies to new realities, and the unrelenting reality of the pacing problem, which makes it increasingly hard for public policy to keep pace with the rate of technological change.

Those reasons are some of the same ones causing soft law to ascend. "Reinventing government" is a phrase that has been used widely in the past, but in light of these new realities, the need to get serious about the reinvention of governance processes is more urgent than ever. This need particularly applies to fast-evolving sectors in which "there is a growing consensus that traditional government regulation is not sufficient for the oversight of emerging technologies," as Wallach and Marchant argue.[4]

For those reasons, many tech policy scholars and governance experts have begun identifying new models that can help address pressing policy concerns without resorting to build-and-freeze hard-law approaches that are no longer working effectively, or which are inappropriate for newer, fast-moving technologies and sectors for which we hope to see accelerated innovation opportunities.

"Co-regulation" is a related term used to describe the give-and-take between regulators and regulated parties that is often a part of soft-law processes.[5] Phil Weiser describes the co-regulation approach as one in which an agency integrates "its efforts with private bodies with expertise in the field" and in which "integration involves the explicit embrace, oversight, and in which enforcement of actions by private bodies" to solve difficult problems outside traditional regulatory processes.[6]

An even broader term for these new approaches is "flexible regulation." A diverse body of scholarship has developed over the past quarter century that outlines flexible approaches to governance in various contexts. In her book *Innovation and the State*, Cristie Ford notes that these models go by many different names—reflexive law, management-based regulation, experimental governance, principles-based regulation, meta-regulation, and others—but that they all share a commitment to move away from the overly rigid and inefficient regulatory methods of the past.[7]

The consultancy firm Deloitte has produced several important reports on the need to reinvent governance frameworks for emerging technologies using flexible or co-regulatory approaches. It argues that these methods are essential today because "[i]nnovative technologies and new business models can catch regulators by surprise."[8] Deloitte recommends a set of five approaches to guide the future of emerging technology policy:[9]

- *Adaptive regulation*: Shift from "regulate and forget" to a responsive, iterative approach.
- *Regulatory sandboxes*: Prototype and test new approaches by creating sandboxes and accelerators, which are mechanisms that allow regulators to experiment with alternative and more flexible governance schemes without having to abandon laws or regulations altogether.
- *Outcome-based regulation*: Focus on results and performance rather than form.
- *Risk-weighted regulation*: Move from one-size-fits-all regulation to a data-driven, segmented approach.
- *Collaborative regulation*: Align regulation nationally and internationally by engaging a broader set of players across the ecosystem.

Many regulatory agencies are already using those approaches to cope with the pace of technological change and address thorny governance issues. But government agencies are not the only ones.

In a 2019 law article with Ryan Hagemann and Jennifer Huddleston, I cataloged a long list of soft-law methods and various case studies describing how such mechanisms are being tapped by government bodies today to deal with fast-moving technologies.[10] A partial inventory of soft-law methods includes multistakeholder processes, industry best practices or codes of conduct, technical standards, private certifications, agency workshops and guidance documents, informal negotiations, and education and awareness efforts. Again, this list of soft-law mechanisms is amorphous and ever-changing. Moreover, many of those soft-law methods and processes are used in conjunction with hard-law methods.

Multistakeholder processes are a particularly important type of co-regulatory soft law, and they have been the cornerstone of America's digital economy policy efforts for two decades.[11] In July 1997, the Clinton administration released *The Framework for Global Electronic Commerce*, a statement of the administration's principles and policy objectives toward the internet.[12] The document said that "governments should encourage industry self-regulation and private sector leadership where possible" and "avoid undue restrictions on electronic commerce."[13] The co-regulatory multistakeholder model promoted by the *Framework* was instrumental in helping transition internet governance and policymaking efforts from the National Science Foundation to the National Telecommunications and Information Administration (NTIA) and the Internet Corporation of Assigned Names and Numbers.[14] That collaborative governance vision has been the cornerstone of internet policy ever since.

Through collaborative efforts, regulators have been working with innovators and various civil society organizations to formalize "privacy-by-design," "safety-by-design," and "security-by-design" efforts.[15] Through ongoing conferences, meetings, negotiations, and guidance documents, these parties have hammered out best practices that bake important values and safeguards directly into the product design process. This work is a way of introducing what some call anticipatory ethics into the early stages of technological developments.[16]

For example, over the past two decades, soft-law mechanisms have been used extensively to address concerns about online safety and youth activities on the internet. Between 2000 and 2010 alone, six major online safety task forces or blue-ribbon commissions were formed to study online safety issues and consider what should be done to address them.[17] Three of those task forces were convened by the U.S. government, and the British government commissioned another. Two additional task forces were formed through universities and private associations during this period. Each of those six task forces was made up of, or received input from, a diverse set of experts from academia and think tanks, corporations and professional trade associations, advocacy organizations, and various government agencies. In other words, they were multistakeholder processes. The task forces recommended a variety of best practices, educational approaches, and technological empowerment solutions to address various safety concerns.

More recently, multistakeholder processes have formulated privacy, safety, and cybersecurity-related best practices. Many of the meetings were convened by the U.S. Department of Commerce (the NTIA in particular); the White House Office of Science and Technology Policy; and a wide variety of federal regulatory agencies, including the FTC, FDA, FAA, and FCC.

Those multistakeholder efforts and agency best-practice reports have contained assorted "responsible innovation" principles for technologies as wide ranging as the following:

- Big data, machine learning, and artificial intelligence[18]
- The Internet of Things (i.e., internet-enabled devices and applications)[19]
- Online advertising practices[20]
- Autonomous vehicles policy[21]
- Motor vehicle cybersecurity[22]
- Cybersecurity of advanced medical devices[23]
- Facial recognition technologies[24]
- Health and medical smartphone applications[25]
- Medical advertising on social media platforms[26]
- Mobile phone privacy disclosures[27] and mobile applications for children[28]
- 3D-printed medical devices[29]
- Small unmanned aircraft systems (i.e., drones)[30]

This list just scratches the surface of soft-law and multistakeholder processes. Moreover, the recommendations flowing out of those soft-law efforts can be quite detailed and are too numerous and context-specific to itemize here. But to illustrate, one common best practice recommended in many of those efforts involves devising appropriate data collection and storage procedures. As part of various soft-law efforts, innovators are typically encouraged to use commonly accepted encryption techniques and ensure that data are properly handled, used only for clearly specified and sensible purposes, and then deleted after a certain period of time. In some cases, technical specifications and procedures are worked out during multistakeholder negotiations. In other cases, those tasks are left to industry bodies or third-party accreditors to address and enforce.

The bottom line is that, just as software is eating the world, soft law is now eating the world of technological governance.[31] Strangely, however, many tech critics or responsible research advocates rarely mention such efforts in their writings. Perhaps they are simply unaware of the many ways in which the principles they advocate already infuse multistakeholder processes and soft-law efforts. But that seems unlikely because those efforts are widely discussed and reported. The more likely reason they ignore soft-law efforts is that they probably do not believe that those efforts are comprehensive or stringent enough. To the extent that they discuss soft-law efforts at all, tech critics or responsible research and innovation (RRI) scholars will often say that such initiatives lack teeth and that anything short of a full-blown regulatory regime (or significant expansion of an existing one) is insufficient to address their concerns. It seems clear that, for many tech critics, only a comprehensive federal law and corresponding regulatory regime for each emerging technology sector will be enough.

That perspective is unfortunate, and if it is the line in the sand that RRI scholars wish to draw, then they will be left with little wiggle room in conversations about governance options for emerging technologies. The prospects for comprehensive regulation of most emerging technologies are dim; at the very least, comprehensive regulation will take many years to get in place. Again, the unrelenting pace of technological change means the clock is always ticking. In many cases, the law would be outdated by the time it got on the books. Practically speaking, therefore, it is a mistake to make the perfect the enemy of the good, and it would be wise to have backup governance plans that are more adaptable in this era of rapid technological change. In this sense, soft-law efforts might be viewed as the minimum necessary governance needed to address various social needs and values. If RRI principles already infuse soft-law governance processes, and if they can be improved in a flexible way to adapt to new challenges, that means a rough-and-ready set of principles and policies are in place while more formal rules are pursued, if they continue to be needed at all.

Some RRI scholars, and potentially many innovators, will decry the messy, uncertain nature of soft-law governance processes. For many such scholars and tech critics, soft law is simply not enough. But there is a great deal to be said for the way soft-law mechanisms have already started adapting in a dynamic, iterative fashion to deal with rapid changes in various fields. Additionally, although soft-law systems may embody various uncertainties, they still often provide *more* governance than hard law. As Chapter 4 documented, hard-law systems regularly struggle to adapt to changing technological realities, even though that reluctance to change creates serious problems and makes those governance systems less effective (and potentially more likely to be evaded by innovators). If soft law offers a better chance than hard law of getting some principles and values baked into technological design processes, then RRI supporters should acknowledge that potential benefit and build on it.

Finally, it is vital to understand that soft law does not develop in a vacuum. There is not an either-or choice between soft law and hard law so much as there is a constantly sliding scale of governance options that ideally are used in cooperation with each other. In most soft-law schemes, government officials initiate the process or are at least a major part of it. Moreover, soft law often builds upon, or operates alongside, many other governance mechanisms, including many that are reactive and remedial in character. These mechanisms include the following:

- Federal and state consumer protection agencies (such as the FTC), which police unfair and deceptive practices and other harms
- Courts and common law, including legal solutions like product liability, negligence, design defects law, failure to warn, breach of warranty, contract law, and other assorted torts and class action claims
- Insurance markets, which serve as risk calibrators and correctional mechanisms
- Third-party accreditation and standard-setting bodies, discussed later
- Education and awareness efforts, both by government bodies and third parties, also as discussed later
- Social norms and reputational effects, especially the growing importance of reputational feedback mechanisms[32]
- Media, academic institutions, nonprofit advocacy groups, and the general public, all of which can put pressure on technology developers
- New entry and competition combined with the power of consumer choice

Only by taking into account the full range of players and activities at work can we develop a more robust understanding of how technology is actually governed in our modern world. I suspect that many RRI scholars *do* appreciate these other factors, even though they sometimes fail to account for all of them in their writing and advocacy. But, again, many of those advocates generally do not favor the remedial, ex post nature of some of these governance tools and will continue to insist that more ex ante anticipatory planning must be at the heart of technological design and development processes.

In reality, a mix of these two approaches is already at work today in soft-law processes and will likely continue to dominate governance well into the future. As long as anticipatory efforts do not become formal regulatory proposals, this mix of responsible innovation governance tools and methods should be embraced by a diverse array of scholars and innovators alike.

Risk Education

Educational approaches are a particularly important part of the soft-law toolkit, yet they are often underappreciated. With traditional regulatory approaches being strained by new realities, public awareness campaigns and risk communication efforts can be an effective way of providing citizens with better information about some of the risks associated with new technologies they increasingly need and demand. Improved risk education can also help address the problem of technological illiteracy that can fuel the sort of technopanics discussed throughout this book.

Social scientists frequently debate the degree to which scientific or technological illiteracy among the general public ends up driving poor decisionmaking—both by individuals and policymakers. The *knowledge deficit model* holds that public skepticism about science or hostility to certain technologies is related to the level of public ignorance or misunderstanding of the technology at hand. In theory, better education about such matters should correct that ignorance and reduce opposition to science and technology.

Other social scientists and economists contend that a certain degree of rational ignorance about new scientific or technological developments exists. Because individuals are limited in time and ability to process highly technical matters, they rely on cognitive shortcuts or trusted sources to inform their attitudes and decisions about science and technology. That model is sometimes referred to as the low-information rationality or bounded rationality model. In either case, improved risk communication and technological literacy efforts can help better inform individuals about complex science and technology issues.

Consider the GMO example discussed earlier. When public understanding of risk tradeoffs is based on myths or misperceptions, it can result in backlashes and technopanic-based policymaking about genetic modification.[33] Better education and risk communications can help reverse that problem.

Regulators can play an important role in this regard as public risk educators. Again, consider digital technologies and online safety. As noted earlier, soft-law mechanisms have been tapped repeatedly to deal with various concerns related to online safety, harassment, and hate speech. The many task forces and blue-ribbon commissions that were organized to address these issues generally agreed that educational approaches would be both more effective and less restrictive than regulatory solutions.[34] The accelerating pace of technological change was a primary factor cited by most of the task forces when reaching that conclusion. More specifically, the task forces outlined how a combination of media literacy, awareness-building efforts, public service announcements, targeted intervention techniques, and better mentoring and parenting strategies could help prepare youngsters to be better digital citizens and to better adapt to changes in technology than could top-down regulations. Many government agencies, such as the FTC and FCC, already work together and with technology developers to facilitate education and awareness efforts about online safety and security threats and best practices. This model is a good one for many other fast-moving, hard-to-classify technologies such as the Internet of Things and artificial intelligence.

Another example of how education can be helpful in communicating risks involves the Food and Drug Administration. Risk communication and health literacy are already important parts of the FDA's mission, even though they do not receive much attention and the agency does not allocate nearly as much resources to them compared to traditional regulatory responsibilities.[35] Risk regulation has always been the primary focus of FDA efforts to ensure safe and effective drugs and medical devices. It will likely remain that way. But as noted in earlier chapters, the old build-and-freeze model of regulation is increasingly under strain from new realities.

Stepped-up risk education efforts can help fill the emerging gaps. In its 2009 *Strategic Plan for Risk Communication*[36] as well as its 2011 report *Communicating Risks and Benefits: An Evidence-Based User's Guide*,[37] the FDA provided a roadmap for what a more comprehensive risk education campaign would entail. The FDA also engages in various product labeling efforts as well as other public education campaigns and strategies.[38] Yet those efforts have always been secondary for the agency, which has instead focused on trying to preemptively guarantee the safety and efficacy of drugs and devices. And much of the education the FDA does is basically explaining to companies and the public how to comply with its voluminous body of regulation.

A more robust focus on risk education would aim to better inform citizens about the relative risk tradeoffs they face with new technologies and technological capabilities.[39] Such risk education should focus on both the general public and the innovators who are providing new devices and treatments. These approaches will be essential in a world of highly personalized medicine, where citizens are more empowered to make their own wellness decisions. As Chapter 2 made clear, new technological capabilities are already giving people more options about how to address their health or augment their abilities using digital health technologies, 3D printers, genetic technologies, and biohacking techniques. Stepped-up risk education and health literacy are desperately needed as these capabilities accelerate and outstrip the ability of traditional laws and regulations to keep up with breaking developments and a more technologically empowered public.[40]

Some skeptics will argue that government will often get things wrong when it engages in risk education or health literacy. For example, many health experts criticize the U.S. Department of Agriculture's Food Pyramid because of specific dietary recommendations that those experts feel have undermined public health.[41] To be sure, government health officials—and officials engaging in risk education in other contexts—will not always get it right. Government-led risk education efforts must pivot and adapt to changing technical and social realities or else they will be ignored. Of course, no one is forced to follow the government's advice. Moreover, governments are not the only ones doing such education. As noted next, many other organizations are helping to advance the understanding of various risks and provide guidelines for acceptable development and use of new technological capabilities.

The Importance of Professional Associations and Ethical Codes

Professional organizations, trade associations, and various consortia can and do develop guidelines and codes of ethics to address the responsible development or appropriate use of various emerging technologies. Such organizations can serve as independent standard-setting bodies and can help hold innovators accountable by designing guidelines and best practices established through soft-law processes.

Various trade associations have already worked with government agencies to formulate some of the best practices and codes of conduct documented earlier. Other organizations have focused on developing high-level codes of professional conduct for innovators in their sectors. Some of the most notable examples involve the Association of Computing Machinery (ACM), the Institute of Electrical and Electronics Engineers (IEEE), the International Organization for Standardization (ISO), and UL (which was previously known as Underwriters Laboratories), among others.

For example, the ACM developed a Code of Ethics and Professional Conduct in the early 1970s, refined it in the early 1990s, and then updated it again just recently in 2018.[42] Each iteration of the ACM Code reflected ongoing technological developments, from the mainframe era to the PC and internet revolution and on through today's machine learning and AI era. The latest version of the ACM Code "affirms an obligation of computing professionals, both individually and collectively, to use their skills for the benefit of society, its members, and the environment surrounding them" and insists that computing professionals "should consider whether the results of their efforts will respect diversity, will be used in socially responsible ways, will meet social needs, and will be broadly accessible."[43] The document also stresses the following:

> An essential aim of computing professionals is to minimize negative consequences of computing, including threats to health, safety, personal security, and privacy. When the interests of multiple groups conflict, the needs of those less advantaged should be given increased attention and priority.[44]

Other organizations formulate more targeted or applied best practices and codes of conduct. The field of artificial intelligence and machine learning is a particularly good example. Several initiatives are already underway:

- The IEEE's Ethically Aligned Design project is an effort to craft "A Vision for Prioritizing Human Wellbeing with Artificial Intelligence and Autonomous Systems."[45] With more than 420,000 members in more than 160 countries, IEEE boasts of being "the world's largest technical professional organization dedicated to advancing technology for the benefit of humanity."[46] IEEE's new effort seeks to incorporate into AI design five key principles that involve the protection of human rights, better well-being metrics, designer accountability, systems transparency, and efforts to minimize misuse of these technologies. The second iteration of the group's report was 263 pages long and contained a litany of recommended best practices to satisfy each of those objectives. The effort included almost a dozen working groups with detailed reports and a variety of certification proposals as well.
- The Partnership on AI began as an industry-led effort formed by Apple, Amazon, Google, Facebook, IBM, and Microsoft, but it has grown to include more than 80 members from industry, civil society organizations, academic institutions, and other groups. The Partnership is billed as a multistakeholder organization that brings those diverse groups together "to study and formulate best practices on AI, to advance the public's understanding of AI, and to provide a platform for open collaboration between all those involved in, and

affected by, the development and deployment of AI technologies."[47]

- OpenAI is a nonprofit research organization created in 2015 with seed money from notable tech innovators and investors like Elon Musk of Tesla (and formerly PayPal), Sam Altman of Y Combinator, venture capitalist Peter Thiel (also formerly of PayPal), Reid Hoffman of LinkedIn, and others. OpenAI publishes research reports discussing how to make sure that AI development "is used for the benefit of all, and to avoid enabling uses of AI or (artificial general intelligence) that harm humanity" and to ensure it does not become "a competitive race without time for adequate safety precautions."[48] OpenAI is also a member of the Partnership on AI.

- In late 2016, the British Standards Institute published a "Guide to the Ethical Design and Application of Robots and Robotic Systems."[49] Developed by a committee of scientists, academics, ethicists, and philosophers, the guide "recognizes that potential ethical hazards arise from the growing number of robots and autonomous systems being used in everyday life" and aims to "eliminate or reduce the risks associated with these ethical hazards to an acceptable level." Specifically, the guide's protective measures create best practices for the safe design and use of robotic applications in a wide range of fields, from industrial services to personal care to medical services.[50]

- The ISO is a global standards–making body that was formed in 1946 and continues to play an important role in establishing international norms for emerging technologies. The ISO "is an independent, non-governmental international organization with a membership of 163 national standards bodies"[51] that seeks to build global consensus through multistakeholder efforts. The ISO uses dozens of technical committees that include global experts in diverse fields, such as industry, consumer associations, academia, nongovernmental organizations, and governments.[52] It has already played an important role in formulating global best practices for robotics and AI-based applications. In 2014, for example, the ISO crafted requirements and guidelines "for the inherently safe design, protective measures, and information for use of personal care robots."[53] That standard is just one of dozens of robotics-related ones that ISO has published.[54]

Other industry groups and professional societies in fields as diverse as drones and biotechnology are developing guidelines and best practices for their sectors. Such efforts often complement governmental efforts to explore issues surrounding emerging technologies.

It goes without saying that codes of conduct, voluntary standards, and professional ethical codes are not cure-alls for the problems associated with technological development, and they cannot magically ensure that all innovation will be responsible. Additional efforts will sometimes be needed, as will be discussed later. But efforts such as those described here can go a long way toward improving accountability and responsibility among various emerging technology companies and individual innovators. Standards, codes, ethical guidelines, and multistakeholder collaborations create powerful social norms and expectations that are often equally important as, or even *more* important than, what laws and regulations might seek to accomplish.[55] There are powerful reputational factors at work in every sector that—when combined with efforts such as these—create a baseline of accepted practice. These efforts are also likely to get more initial buy-in among private innovators, at least compared to heavy-handed regulatory proposals. Finally, these efforts deserve more attention if for no other reason than the continuing reality of the pacing problem. Soft-law mechanisms will always be easier to adopt and adapt as new circumstances demand.

Critics might also insist that, even if they do *some* good, privately negotiated best practices or codes of conduct will be hard to coordinate and enforce both domestically and abroad. This concern is valid, but it may be intractable in light of the different values that various countries and cultures possess.

To help address this problem, Gary Marchant and Wendell Wallach propose the formation of what they call governance coordinating committees (GCCs). GCCs would help coordinate technological governance efforts among governments, industry, civil society organizations, and other interested stakeholders in fast-moving emerging technology sectors.[56] Because "no single entity is capable of fully governing any of these multifaceted and rapidly developing fields and the innovative tools and techniques they produce," they suggest that GCCs could act as a sort of "issue manager" or "orchestra conductor" that would "attempt to harmonize and integrate the various governance approaches that have been implemented or proposed."[57] They have also called for the formation of an International Congress for the Governance of AI as "a first step in multistakeholder engagement over the challenges arising from these new technological fields."[58]

In essence, Marchant and Wallach are proposing the creation of what is commonly known in Europe as a *quango*, or quasi-autonomous nongovernmental organization. Quangos have been effective in Europe and some other areas in helping devise solutions to governance coordination challenges in technically complicated fields. Like quangos, GCCs could help provide another mechanism whereby technological governance issues are addressed through ongoing collaboration among various parties, both domestically and globally. They could help craft or enforce voluntary best practices, or at least offer a forum for ongoing discussions around thorny issues. Difficult details and hard questions remain, including the following: how do GCCs get formed, who would be on them, and how would they be supported financially? Moreover, even to the extent that international consensus could be found on ethically complicated issues (like genetic modification of human embryos), how would a GCC be able to enforce restrictions across so many countries and cultures?[59] As a forum for conversation and collaboration, however, GCCs could still hold great promise and should be given greater consideration.

The Challenge of Defining Harm

Soft law cannot serve as a complete substitute for hard law. Some technological developments can give rise to significant harms or intractable problems that will sometimes require a heightened level of regulatory scrutiny and action. That fact does not mean hard-law solutions will always be completely effective in solving those problems, and at times hard law may create more problems in the process. Nonetheless, laws and regulations will sometimes need to be considered to help discourage the most serious harms associated with certain technological developments.

How are policymakers supposed to determine which technologies and theoretical harms deserve more regulatory scrutiny, and how should they determine the likelihood or measure the potential severity of the harms? These are notoriously hard questions to answer because there are many different ways to judge what constitutes acceptable risk[60] or catastrophic risk.[61] Such questions also demand a fuller exploration of what theories of rights and responsibilities animate the discussion. Here I offer only a brief sketch of a much-needed theory of technological harm that can help us answer these questions. But even a robust theory would not be able to preemptively answer every question critics pose about the alleged dangers of various emerging technologies. A major point of my previous book was that many, if not most, of these questions can only be answered through real-world, trial-and-error experiences and responses.[62]

To understand why that is the case, again consider the many contentious debates about online safety and privacy. Many innovation policy squabbles—both today and in the past—have involved heated battles over what are best thought of as cognitive or psychological harms. Those harms are not to physical life and limb but rather are potential harms to one's feelings or cognitive processes or the creation of a general sense of unease.

Supposed informational harms were also at the heart of past policy skirmishes over indecent or obscene content. Raging debates surrounded these issues long before the internet came along, and policymakers imposed many censorial prohibitions on content creators and distributors in the name of upholding "community standards" and "public decency."[63] To be sure, a great many people thought that there was some harm to themselves, their children, or the public more generally because of such content. Yet large numbers of people disagreed with the proposed regulations and felt it was their right to consume whatever sort of content they desired. How should harm be calibrated when "objectionable content" is in the eye of the beholder? In this case, the First Amendment generally won out over time, and most content controls gradually went away. Even though some rules are still on the books today, they are largely ignored as agencies and courts engage in the sort of rule departure discussed earlier. In essence, we have witnessed the end of traditional censorship efforts in the United States over the past two decades.[64]

Eye-of-the-beholder spats and calls for information controls have not gone away, however. Instead, they have moved into new areas. For example, subjective theories of informational harm have become a major fault line in debates over digital security and privacy. Tech critics often insist that new data collection and dissemination capabilities pose a threat to their security and privacy "rights." Others do not seem to understand what all the fuss is about and worry more about how their regulation might impede their "right" to collect and receive more information or speech, or perhaps their "right" to better service, greater convenience, or lower prices for important services. Whose rights should prevail, and are these really rights at all? In 2017, the Federal Trade Commission even launched an inquiry into the question of what constituted informational harms in the context of online data collection or data security incidents.[65] Among the filings that the agency received in this and related proceedings, little consensus existed regarding what constituted information rights or wrongs.[66]

These debates might never cease, because we might never be able to reach strong consensus regarding the nature and extent of such rights or determine how to protect those values.[67] Moreover, how those rights are conceived of varies widely by country, making global enforcement more challenging. Many European laws conceive of privacy as a "dignity right" that trumps most other economic and social values, including freedom of speech. The United States has taken a different approach. Privacy rights have been generally associated with other, more well-established rights that are more tangible in character, including the property rights people hold in their bodies, their homes, their personal property, and their financial accounts. To the extent that there is an overarching information imperative in the United States, it has been shaped by the First Amendment to the Constitution, which generally disallows regulation of the collection or use of data with a few important exceptions (personal health and financial information, for example).

Privacy advocates in the United States insist that policymakers should mimic the European approach and adopt a comprehensive privacy law that might even incorporate a formal "privacy bill of rights." Throughout the Obama administration, privacy advocates pushed such efforts but failed to get any traction. At the time of this writing, however, America appears on its way to potentially advancing a new federal privacy framework, if for no other reason than to preempt a confusing patchwork of state privacy laws. Even if these measures pass, however, they will be severely challenged by all the same realities documented throughout this book. Government passage of major bills claiming to protect privacy does not necessarily mean such laws accomplish the goal envisioned. Technologies, individual desires, and societal values continue to evolve faster than laws in many instances. A great many Americans enjoy the benefits associated with data collection, including lower-cost services, various conveniences, expanded competition and choice, and other benefits, and they would likely choose to continue to share their data in exchange for those things. Sometimes social values and individual choices are as hard to control as technological change.

Soft law is preferable to hard law when consensus is elusive, as it is in many of these cases. An adaptive multistakeholder framework performs better than codified laws when harms are amorphous, speculative, or subjective. This reason is partially why various soft-law processes are being tapped more regularly. When coupled with ex post judicial remedies, soft law continues to represent the better approach for most online safety, privacy, and security concerns, as well as a variety of other concerns about emerging technologies. The threat of hard law can sometimes help discourage the worst types of misbehavior by some actors, or at least encourage them to come to the table as part of a multistakeholder process and agree to commonly accepted best practices going forward.

When Soft Law Isn't Enough: Existential Risks

What about more serious alleged harms where widespread agreement exists that more should be done to preemptively address risks to life, limb, health, and so on? The most problematic category of such harms is often referred to as "existential" or "catastrophic" risks. As noted in Chapter 1, "existential" is a term some tech critics throw around far too casually when decrying a variety of innovations or particular companies they do not care for. We can dismiss assertions of existential threats when they are alleged for lesser matters, such as whether Facebook is destroying civilization as we know it. That sort of threat inflation cheapens the meaning of the term "existential"; there are no plausible mechanisms by which Facebook could pose such a threat. There may be legitimate existential threats out there that we *should* be spending more time addressing, but that threat probably isn't one of them.[68]

Nick Bostrom, Director of the Future of Humanity Institute at the University of Oxford, has written extensively about the dangers of "superintelligence" and what he calls the "vulnerable world hypothesis." What makes Bostrom's work distinctive among modern technology critics is his willingness to finish his sentences. That is, critics usually heap scorn on various technologies, but most do not follow through with concrete recommendations for what to do about their litany of woes. By contrast,

Bostrom provides a roadmap with various options about how to address the new technological risks he believes exist. This roadmap makes Bostrom's work deserving of greater attention because it signals what sort of regulatory approaches other critics and policymakers might eventually support.

"Our approach to existential risks cannot be one of trial-and-error," Bostrom argues, because with such risks, "[t]here is no opportunity to learn from errors."[69] In other words, some theoretical risks are so potentially catastrophic that permissionless innovation is no longer the optimal default for tech policy. Does that automatically mean that the precautionary principle should be the default? Not necessarily. As Bostrom himself notes, "stopping technological development would require something close to a cessation of inventive activity everywhere in the world. That is hardly realistic; and if it could be done, it would be extremely costly—to the point of constituting an existential catastrophe in its own right."[70]

On the other hand, Bostrom argues, "*limited* curtailments of inventive activities" might be a sensible policy.[71] That approach was adopted by governments to address the use of chemical weapons after World War I, and then nuclear proliferation after World War II. After the horrific uses of chemical weapons during World War I, the Geneva Protocol for the Prohibition of the Use in War of Asphyxiating, Poisonous or Other Gases, and of Bacteriological Methods of Warfare was formulated in 1925 to limit the uses of such weapons in future conflicts.[72] Later, after World War II, international treaties and other agreements were formulated that sought to limit the ability to possess or enrich uranium, or to traffic nuclear weapons. The Treaty on the Non-Proliferation of Nuclear Weapons (NPT) was created in 1968 to advance the peaceful uses of nuclear technology while seeking to limit the dangerous ones. The International Atomic Energy Agency (IAEA), formed a decade earlier, helps advance this mission "to accelerate and enlarge the contribution of atomic energy to peace, health and prosperity throughout the world."[73]

This book is not the place for a comprehensive evaluation of the success of the Geneva Protocol for chemical weapons or the NPT and the IAEA for nuclear proliferation. Nonetheless, we can draw two high-level conclusions from these efforts. First, the worst fears about chemical and nuclear weapons have not come to pass. Although rogue actors still exist and develop such weapons, the most concerning applications of these technologies have been constrained for the most part. But how much of that success can be attributed to treaties and nonproliferation agreements versus the simple fact that it is costly to obtain and produce such weapons? If the cost and complexity of weaponization *were* the primary factors limiting the worst-case applications of chemical and nuclear technologies, what happens in a world in which newer dangerous technologies are cheaper, more widely available, and easier for greater numbers of people to access or develop?

That fear leads Bostrom to propose the "Principle of Differential Technological Development," which would

> Retard the development of dangerous and harmful technologies, especially ones that raise the level of existential risk; and accelerate the development of beneficial technologies, especially those that reduce the existential risks posed by nature or by other technologies.[74]

Although Bostrom admits that "correctly implementing differential technological development is a difficult strategic task," he believes "it is worth making the attempt" if for no other reason than "to buy a little time."[75] Alas, his proposed specific measures and countermeasures to mitigate vulnerabilities all have rather serious tradeoffs and limitations. Bostrom recommends that we consider efforts that would do the following:

- Prevent dangerous information from spreading.
- Restrict access to requisite materials, instruments, and infrastructure.
- Deter potential evildoers by increasing the chance of their getting caught.
- Be more cautious and do more risk assessment work.
- Establish some kind of surveillance and enforcement mechanism that would make it possible to interdict attempts to carry out a destructive act.

In thinking about enforcement options, one must consider both the practicality and the wisdom of each approach. To move away from the theoretical and toward the practical, one can apply Bostrom's framework to modern existential threats that are commonly discussed, such as concerns about 3D-printed weapons or the development of robotics and AI technologies. Much like chemical and nuclear technologies before them, 3D printers, robots, and AI technologies already have many important peaceful and socially beneficial uses, and many more are sure to be developed. But, if used improperly, these technologies could produce horrific consequences. As Chapter 1 already noted, all one needs to do is read or watch just about any sci-fi book or show about robots or AI to see every worst-case scenario explored *ad nauseam*.

So, what should be done to prevent the rise of 3D-printed "ghost guns," "killer robots," and *Terminator*-esque scenarios? Although far-fetched, such occurrences are, at least, risks that might warrant some degree of precautionary regulation.

Returning to Bostrom's proposals for dealing with existential risks of this sort, the first of them—restricting the spread of dangerous information about new technologies or technological capabilities—is probably the least feasible in a world of ubiquitous, low-cost information transmission. In addition, it is not wise to propose a global censorship regime in this regard because of the potential collateral damage it would have for beneficial types of information flows.

Bostrom's last suggestion—preventive policing through stronger interventions by various levels of government—raises new risks of its own. "If the continued survival of humanity depended on successfully imposing worldwide surveillance," responds Kelsey Piper of *Vox*, "I would expect the effort to lead to disastrous unintended consequences."[76] A mass surveillance apparatus would not necessarily guarantee workable containment solutions to the sort of disasters that Bostrom fears, but it certainly would open the door to a different type of disaster in the form of highly repressive state controls on communications, individual movement, and other activities. Even assuming we could look beyond the specter of mass surveillance in the name of reducing technological risk, the questions of cost and resource constraints remain a problem. Bostrom does not consider those downsides, however.[77] Finally, mass surveillance schemes could discourage research into a great many risk-*reducing* technological applications and thus undermine Bostrom's other goal of "accelerat[ing] the development of beneficial technologies, especially those that reduce the existential risks posed by nature or by other technologies."

That leaves his three other options, all of which have greater merit. To repeat, those options are as follow:

- Restrict access to requisite materials, instruments, and infrastructure.
- Deter potential evildoers by increasing the chance of their getting caught.
- Be more cautious and do more risk assessment work.

These recommendations more closely track the approaches and instruments that were developed to deal with high-risk uses of chemical and nuclear weapons. In fact, since 2012, there has been a major effort underway called the Campaign to Stop Killer Robots, which seeks a multinational treaty to stop the most nefarious robotic applications.[78] At the time of this writing, almost 30 countries, 86 nongovernmental organizations, and more than 25,000 AI experts had pledged support of this effort "to ban fully autonomous weapons and thereby retain meaningful human control over the use of force." Meanwhile, almost 250 organizations and more than 3,000 individual experts have signed the Future of Life Institute's "Lethal Autonomous Weapons Pledge," which "call[s] upon governments and government leaders to create a future with strong international norms, regulations and laws against lethal autonomous weapons."[79] Signatories vow that they "will neither participate in nor support the development, manufacture, trade, or use of lethal autonomous weapons."[80]

It remains unclear how enforcement will work, but one could imagine that "killer robot" applications might be limited through international accords and actions, perhaps using the Geneva Protocol for chemical weapons or the NPT and the IAEA for nuclear proliferation as models. A similar framework might be considered for 3D-printed weapons or even certain synthetic biology or genetic engineering applications that involve extreme forms of human modification. The UN's Biological Weapons Convention framework might provide a model in these cases. The International Criminal Court, whose mission is to "to hold those responsible accountable for their crimes and to help prevent these crimes from happening again," could also play a role in addressing lethal uses of emerging technologies.[81] Over time, the body of laws, accords, and general principles that make up the law of armed conflicts will evolve to accommodate these new technological capabilities.

To be sure, such approaches are not foolproof. How should we deal with rogue states or other holdouts who refuse to play a constructive role in such agreements and treaties? We already face that problem with nuclear nonproliferation efforts and states like North Korea. More problematic is the question we have already alluded to regarding the regulation of dual-use technologies. Namely, how can we address the harmful applications of various general-purpose technologies (computing, robotics, 3D printers, genetic editing,[82] etc.) without undermining the many beneficial and life-enriching applications of those same technologies?[83]

In this regard, global regulation of genetic editing will soon become an important test case. In March 2019, several of the world's leading genetic scientists came together and called a worldwide five-year moratorium on DNA editing for purposes of producing genetically modified children.[84] The scientists asked governments to "publicly declare that they will not permit any clinical use of human germline editing for an initial period [of five years]."[85] Interestingly, many other top geneticists refused to sign the call for a ban, even though it was just a request for a voluntary moratorium, not a formal treaty. Those not signing the call for a moratorium cited a variety of factors, including the fact that it seemed to be too late for such a ban to be meaningful, with the proverbial genie already well out of the bottle.[86] The scientists agreed that there were serious ethical issues surrounding genetically edited children, but consensus proved elusive about the regulatory specifics. There were unanswered questions about who would enforce the moratorium and how they would do so, especially against rogue actors operating in states that will not honor such a ban.

In a world where innovation arbitrage is only getting easier, dual-use technologies will be harder to control because they and their creators will, as Richard Posner has noted, "simply gravitate to another country."[87] That observation is particularly true today because *physicality* matters less than it did in the past. In a world of ubiquitous and near-instantaneous information flows, how can we really control the ultimate threat: the spread of knowledge about dangerous ideas and applications? But that still leaves open the wisdom and practicality of regulating dual-use technologies more generally. If not properly targeted and limited in nature and scope, overzealous bans on broad classes of technologies could undermine scientific discovery and the many accompanying life-enriching and life-saving benefits specific technologies could bring about.

Risk Prioritization Is Essential

To reiterate, these questions are extraordinarily challenging, with no easy answers. Even though I have repeatedly stressed the benefits of allowing most innovation to develop relatively unencumbered, there will always need to be some limits on those technologies that have the potential to bring about more serious risks to humanity. Precautionary restraints are most justifiable when the alleged harms are highly probable, tangible, immediate, irreversible, catastrophic, or directly threatening to life and limb in some fashion.[88] The argument for permissionless innovation as a *general* default should not be viewed as a demand for unfettered freedom to innovate in *every* instance. That was obviously the case for nuclear and chemical weapons, and it is why Bostrom and others are correct to raise questions about future technological developments that could produce similar existential risks to civilization. We need some prior restraints on technological innovation in such instances.

But, again, perspective is essential. The three important things to remember about technological risk are the following:

- Not all technological risks are equal.
- Almost all technological risks have corresponding *rewards* that must also be considered (or, stated differently, there can be no reward without risk taking).[89]
- Knowledge and resource constraints challenge our ability to predict the course of technological developments.

Because of all these factors, it is vital to weigh the full range of tradeoffs associated with any proposed solution(s) to alleged technological risks. The most important thing that policymakers can do in this regard is to get smarter about risk prioritization and to stop making risks seem greater than they are. Debates about technological risk are haunted by false equivalence in technological risk assessment. Chapter 1 offered several examples of technology critics resorting to false equivalence and threat inflation when discussing tech policy issues. Those views can lead to a paradox in that society might spend so much time and energy panicking over lesser risks that it fails to properly address the ones that are truly significant.[90] In other words, it is the proverbial boy who cried wolf.[91]

Consider, for example, a 2016 address by the then–UN secretary-general Ban Ki-Moon on the Non-Proliferation of Weapons of Mass Destruction (WMDs).[92] In his remarks, the secretary-general advocated a stepped-up disarmament agenda "to prevent the human, environmental and existential destruction these weapons can cause."[93] Ban rightly pressed the need to remain vigilant in addressing the horrors of chemical, biological, and nuclear attacks. He did not stop there, however. The secretary-general went on to discuss his concerns about "new global threats emerging from the misuse of science and technology, and the power of globalization."[94] His speech included a diverse class of emerging technologies that are not usually mentioned in the same breath as those traditional WMDs, including information and communication technologies (ICTs), artificial intelligence, 3D printing, and synthetic biology. Ban said such technologies "will bring profound changes to our everyday lives and benefits to millions of people" but worried that "their potential for misuse could also bring destruction. The nexus between these emerging technologies and WMD needs close examination and action."[95]

There is nothing wrong with Ban raising concerns about many of these emerging technologies.[96] Yet by so casually moving from a heated discussion of traditional WMDs into a brief discussion about the potential risks associated with ICTs, AI, 3D printing, and synthetic biology, he implies that these technologies and their potential risks are roughly equivalent. But it is simply not the case that all these risks are equal. The secretary-general is using what rhetoricians refer to as an appeal to fear. Douglas Walton, author of *Fundamentals of Critical Argumentation*, outlines the argumentation scheme for fear-appeal arguments as follows:[97]

- *Fearful Situational Premise*: Here is a situation that is fearful to you.
- *Conditional Premise*: If you carry out A, then the negative consequences portrayed in the fearful situation will happen to you.
- *Conclusion*: You should not carry out A.

This logic pattern is known as *argumentum in terrorem* or *argumentum ad metum*.[98] Tech critics and other concerned parties sometimes use fear appeals in an attempt to shake the public or policymakers out of a perceived slumber and get them to pay more attention to new technological risks. The problem with fear appeals, however, is that they are often logical fallacies built on poor risk analysis or even outright myths.[99] Yet if such appeals are successful, they can lead to unnecessary anticipatory regulation of emerging technologies.

Ban's speech presents that problem. When important international officials like Ban group all these technologies together in a speech about weapons of mass destruction and sandwich them between impassioned opening and closing statements about the need "to take action" because "the stakes are simply too high to ignore," we are witnessing a fear appeal in action. The conclusion that follows from such appeals is obvious: global controls of some sort are needed. Again, this is a false equivalence. Ban's mistake is to equate all these technologies and risks and then suggest that sweeping action is needed for all of them when, in reality, such actions are probably only appropriate for a smaller class of technologies that legitimately pose an existential risk to humanity.

Policymakers and international figures of importance should be extremely cautious about the language they use to describe new classes of technologies, lest they cast too wide a net. Suggestions that every new technology poses a catastrophic or existential risk will desensitize people to actual risks that may be associated with a narrower class of innovations. That does not mean we should ignore risks associated with other technologies or technological capabilities. Instead of using fear appeals and advocating extreme (and likely unworkable) global regulatory schemes, however, it will often be wiser to build on existing laws, norms, and alternative governance frameworks. It is important to be practical. It most contexts, it remains highly unlikely that a global governance solution will work. We are not likely to witness the development of strict global laws and regulatory bodies, at least not any with serious teeth. The better role that international bodies and actors can play is as coordinators of national policies and conveners of ongoing deliberation about multinational concerns.[100]

For example, a variety of transparency laws and other efforts already exist in many national and global governance regimes. These include know-your-customer guidelines and whistleblower processes that aim to identify problematic actors in various contexts. More resources could be plowed into such efforts. Again, education and awareness-building efforts can also be tapped in many cases. Soft law still has a role to play in this regard, too. Even if we cannot achieve global consensus on the potential harms associated with particular technologies or figure out how to successfully craft a formal global regulatory regime to address those concerns, less formal governance efforts can still help create important ethical norms. Best practices or codes of conduct for researchers and developers can also go a long way toward fostering a culture of responsibility and a greater commitment to safety, as even Bostrom has acknowledged.[101] These options should at least be given a greater chance to help start a conversation about wise technological development and responsible innovation.[102]

Finally, Marchant and Wallach's GCCs idea, discussed earlier, might have some merit in this regard—assuming we can figure out how to create them and make them work in various contexts. In the field of digital communications coordination and internet domain name management, the Internet Society (founded in 1992) and the Internet Corporation of Assigned Names and Numbers (ICANN) are examples of governance coordinating committees of sorts.[103] But ICANN deals with more technical matters that do not involve existential risks. Consensus and coordination will likely prove more challenging in the same areas where it is potentially most needed. If, however, policymakers can get risk priorities right and zero in on the most serious harms, it at least gives society a chance to better address those issues in a rational fashion while allowing other important innovations to develop freely.

8

DEFENDING INNOVATION: A BLUEPRINT

The two previous chapters considered possible frameworks for controlling the pace and nature of technological innovation, but this chapter offers a variety of suggestions for how to create a policy environment conducive to ongoing technological change and the freedom to innovate more generally. Because technological innovation—bolstered by sound policies and good institutions—is a crucial determinant of long-term prosperity, bold ideas are needed to break the traditional logjams that hold back life-enriching innovations, scientific progress, and economic growth. Regulation will always be needed to address serious technological risks. But in a great many instances, far too much unnecessary or outdated regulation already exists and needs to be refined or removed.

Special interests seeking to protect their turf are typically the most formidable problem. Individuals and organizations that have a stake in preserving the status quo aggressively push back against change. They will make reform efforts difficult, and because they often control political processes, they will succeed much of the time.

Equally problematic are the many academics who dispute the benefits of technological change or sneer at values like consumer choice, convenience, or cost reductions. No wonder, then, that they advocate adding more friction to technological processes to slow things down to their preferred speed.[1] As we saw in Chapter 6, those critics will often rally under the banner of humanism even though their actions are often at odds with the actual processes of human betterment. Denying humans the ability to develop new and better tools to improve their lives is fundamentally *anti-human*.[2]

Finally, the most problematic enemies of innovation are those politicians who engage in fearmongering or throw a wrench in the wheels of progress to cater to special interests.[3] When politicians let anti-innovation forces commandeer the regulatory apparatus of the modern state, it can limit the potential of a great many life-enriching technologies to see light of day.

The enemies of innovation are powerful and already have created a sprawling regulatory leviathan that encumbers almost every facet of human activity. It will be extraordinarily difficult to undo this mess. Nonetheless, we have to start somewhere. Sometimes policy reforms *are* possible. It takes a unique alignment of factors and forces to achieve reform, but it is worth pursuing. One factor working in favor of reform is the simple fact that the so-called pacing problem sometimes acts more like the pacing *benefit* by forcing policymakers to reevaluate the wisdom of backward policies that long ago ceased to serve the public interest.

It will take a team effort to advance pro-innovation reforms. Educational institutions, innovation advocates, legal defense groups, and consumers themselves can also help expand entrepreneurial opportunities in important ways. The following reform objectives combine these various strategies.

First, one quick note. Many factors affect the innovative capacity of a nation, including tax policy, labor market and immigration policies, research and development issues, intellectual property law, competition policy, and education issues, among others. Although all these factors are important and deserve greater investigation, the ideas discussed here are focused more squarely on identifying direct barriers to entrepreneurial activity and how to overcome them to advance the freedom to innovate.

Nurture an Innovation Culture by Embracing Permissionless Innovation

The most important thing that policymakers can do to nurture a culture conducive to entrepreneurial freedom is to articulate and defend a culture of permissionless innovation in their words and actions.[4] Scholars in many different fields have noted how political attitudes and pronouncements help determine the innovation culture of a nation.[5] The elements of innovation culture include "attitudes towards innovation, technology, exchange of knowledge, entrepreneurial activities, business, uncertainty," and related activities.[6]

The economic historian Joel Mokyr has noted how "technological progress requires above all tolerance toward the unfamiliar and the eccentric" and that the innovation that undergirds economic growth is best viewed as "a fragile and vulnerable plant" that "is highly sensitive to the social and economic environment and can easily be arrested by relatively small external changes."[7] Specifically, societal and political attitudes toward growth, risk taking, and entrepreneurial activities (and failures) are important to the competitive standing of nations and the possibility of long-term prosperity.[8] "How the citizens of any country think about economic growth, and what actions they take in consequence, are," Benjamin Friedman observes, "a matter of far broader importance than we conventionally assume."[9]

For example, former Federal Reserve chairman Alan Greenspan and coauthor Adrian Wooldridge have observed that "[t]he key to America's success lies in its unique toleration for 'creative destruction'" and an "enduring preference for change over stability."[10] This observation is consistent with the findings of Deirdre McCloskey's recent trilogy about the history of modern economic growth. McCloskey meticulously documents how an embrace of "bourgeois virtues" (including, especially, positive attitudes about markets and innovation) was the crucial factor propelling the Industrial Revolution.[11] More recently, positive attitudes toward innovation and risk taking were equally important for the information revolution. The importance of the bourgeois virtues also helps explain why so many U.S.-based tech innovators became global powerhouses while firms from other countries tended to flounder because their innovation culture was more precautionary.[12]

There are limits to how much policymakers can do to influence citizens' attitudes toward innovation, entrepreneurialism, and economic growth. When policymakers set the right tone with a positive attitude toward innovation, however, that tone inevitably infuses various institutions and creates powerful incentives for entrepreneurial efforts to be undertaken. This position, in turn, influences broader societal attitudes and institutions toward innovation and creates a positive feedback loop.[13] "If we learn anything from the history of economic development," argued David Landes in his magisterial *The Wealth and Poverty of Nations: Why Some Are So Rich and Some Are*

So Poor, "it is that culture makes all the difference."[14] Research by other scholars finds that "existing cultural conditions determine whether, when, how and in what form a new innovation will be adopted."[15]

Economists like Mancur Olson speak of the importance of a structure of incentives that helps explain why "the great differences in the wealth of nations are mainly due to differences in the quality of their institutions and economic policies."[16] In this sense, institutions include what Elhanan Helpman defines as "systems of rules, beliefs, and organizations,"[17] including the rule of law and court systems,[18] property rights,[19] contracts, free trade policies and institutions, light-touch regulations and regulatory regimes,[20] the freedom to travel, and various other incentives to invest.

These ideas, incentives, and institutions are mutually reinforcing, and they also influence the competitive standing of nations in the modern global economy.[21] I have already mentioned how the United States made permissionless innovation the foundation of internet policy when the Clinton administration published its *Framework for Global Electronic Commerce* in 1997.[22] The *Framework* was a clean break from the top-down, inflexible regulatory model that had governed traditional communications and media technologies. Instead, the Clinton administration argued that "the Internet should develop as a market driven arena not a regulated industry." To the extent that government intervention was needed at all, the *Framework* insisted that "its aim should be to support and enforce a predictable, minimalist, consistent and simple legal environment for commerce."

The *Framework* is a particularly powerful example of the positive relationship between attitudes, institutions, and incentives and how that relationship fueled an explosion in entrepreneurial activity.[23] Alas, it is rare for the highest-ranking public official in the United States and an entire administration to embrace permissionless innovation as enthusiastically as President Clinton and his administration did in the late 1990s. Nonetheless, President Clinton's vision for digital innovation can and should be encouraged for other sectors and technologies. In recent years, Congress has passed resolutions endorsing similar principles for other technologies. In 2015, for example, the U.S. Senate unanimously approved a bipartisan resolution aimed at encouraging the development of the Internet of Things.[24] The resolution called for the United States to "develop a strategy to incentivize the development of the internet of things in a way that maximizes the promise connected technologies hold to empower consumers, foster future economic growth, and improve our collective social well-being."[25] Importantly, the resolution also held that the government should seek to use Internet of Things technologies "to improve its efficiency and effectiveness and cut waste, fraud, and abuse whenever possible."[26] Statements and resolutions such as those can help ensure that pro-innovation principles become the lodestar of future technology policy decisions in specific sectors.

Remove Barriers to Entry and Innovation and Protect the Right to Earn a Living

Economists and political scientists generally agree that competition and new entry are essential for expanding economic opportunity and enhancing consumer welfare.[27] Unfortunately, many laws and regulations create direct or indirect barriers to the emergence of new ideas and organizations.[28] Removing such barriers will have profoundly positive results for society. As Dustin Chambers and Jonathan Munemo find, "cutting entry-related red tape is generally associated with superior economic outcomes, such as higher per capita income, reduction in the size of the unofficial economy, less corruption, and improvement in productivity."[29] More concretely, eliminating barriers to innovation helps create new jobs, improves worker opportunities and mobility, and gives consumers more and better goods and services.[30]

As detailed in Chapter 4, federal and state licensing regulations often create particularly problematic barriers to entry and innovation. Federal licenses come in many shapes and sizes, but almost all technocratic regulatory bodies (FAA, FCC, FDA, etc.) impose some sort of licensing requirements on specific technologies and service providers. Federal licensing regulations are notoriously difficult to reform because of massive special interest opposition to changing the status quo. A federal regulatory regime is like a giant Christmas tree that has many gifts for many different groups sitting underneath it year-round. Needless to say, once some groups have been promised those gifts, they do not take kindly to the idea of sharing them or having them taken away. Those gifts take many forms—line-of-business restrictions, exclusive geographic service territories, time-limited permits to operate exclusively, technical service specifications, and sometimes also tax credits or direct subsidies. Even if those regulatory measures were well intentioned and meant to serve the public interest, the reality is that they have often undermined the possibility of greater entry and innovation. Broadcast television and radio operators, for example, have benefited from almost all such rules, thanks to decades-old statutes and FCC regulations that help protect licensed broadcasters from the gales of creative destruction that blow through other sectors.

The most obvious way to reform such policies is through comprehensive statutory deregulation and agency downsizing. But those approaches are increasingly nonstarters for the reasons identified in earlier chapters. Such efforts should not be abandoned entirely, but reform advocates will need to get more creative if they hope to reduce the regulatory burden and open up new entry opportunities. In a moment, I will discuss how something called the Innovator's Presumption can help in that regard. Stepped-up judicial efforts will also be needed to defend entrepreneurs, perhaps in the form of an Innovator's Defense Fund. That proposal is also summarized below.

Before discussing those ideas, let's explore the other major category of entry barriers: state and local occupational licenses. Chapter 4 documented how a complex patchwork of licensing and permitting schemes encumbers new entrants in many different fields, with roughly 20 percent of all professions now being licensed in some form (up from just 5 percent in the early 1950s).[31] Rules vary widely from state to state, making compliance particularly challenging for workers who drive across borders or relocate regularly, such as military spouses and immigrants.

At least in theory, state and local occupational licensing and permitting schemes should be somewhat easier to reform than federal licensing schemes, but this rarely proves to be the case in practice. Entrenched interests at the state and local level often have just as much sway over politicians and regulators as they do at the federal level. And those interested are far more organized and vociferous in defending regulations that help them preserve their turf.

Nonetheless, licensing reforms are finally gaining more traction. These reform efforts got a shot in the arm following the U.S. Supreme Court's 2015 decision in *North Carolina State Board of Dental Examiners v. Federal Trade Commission*, which held that local authorities cannot claim broad immunity from federal antitrust laws when they delegate power to "nonsovereign actors" such as licensing boards.[32] Licensing boards are often given broad powers to regulate professions and limit entry according to rules of their own making. This power is problematic because those licensing boards are typically made of individuals who have already been licensed and therefore have an interest in using the rules to defend themselves from new entry or innovation.

The Supreme Court made it clear that as state and local governments continue to delegate power to licensing boards, they will have to be more transparent and accountable about it. Specifically, the Supreme Court has held that public officials must both "clearly articulate" and "actively supervise" licensing arrangements and regulatory bodies if they hope to withstand federal antitrust scrutiny. Federal antitrust laws can be used to address parochial regulations that restrain trade, foster cartelization, or harm consumer welfare in other ways. In reality, however, federal officials only rarely intervene to address these barriers to competition and innovation.

To address barriers to competition, the Federal Trade Commission has pressured state and local governments to reform or eliminate these barriers in past decades.[33] Most recently, in 2017, the FTC created an Economic Liberty Task Force to investigate these burdens, and the agency issued a report 18 months later offering some ideas about how to make licenses more portable.[34] For both constitutional and practical reasons, however, federal officials can only do so much to address state and local barriers to entry and innovation. Reform needs to come from the bottom up.

In the wake of the *North Carolina Dental* decision, some states have made comprehensive licensing reform a priority. States like Nebraska, Arizona, Ohio, and West Virginia have introduced or passed legislation aimed at addressing licensing burdens. Some measures focus on reforming or sunsetting occupational licensing boards, while others would let unlicensed workers operate in most licensed professions as long as they notified customers of that fact. Some of the measures seek directly to protect the right to earn a living.[35] In other states, licensing barriers are sometimes addressed sector by sector, such as reforms that would legalize home-based baking businesses that would then not need a license to sell their baked goods.[36]

Venture capitalist John Chisholm, author of *Unleash Your Inner Company*, has noted that "many potential entrepreneurs . . . are being blocked by regulations. The numbers blocked each decade grow as regulations grow. The men and women in society who find it hardest to provide for themselves and their families and live in self-sufficient dignity are blocked."[37] Chisholm is correct to use the term "dignity" in relation to the idea of being able to freely pursue a living and provide for oneself and loved ones. Making a living is, after all, one of the most essential human pursuits.

Although it would be challenging, complete reciprocity among the states in terms of occupational licenses would be optimal. That is, if workers earn the right to pursue a living in one state, they should be able to do so in others. After all, citizens can use their driver's licenses to operate vehicles in any state, yet a licensed worker needs a new permit in each jurisdiction. That makes no sense. It limits worker mobility, economic opportunity, and workers' rights to earn a living.[38] That sort of reciprocity will be difficult to achieve, however, because coordinating a nationwide pact would face practical and legal challenges. This reality means that state-by-state licensing reform will likely remain the primary option in the near term.

Protect the Innovator's Presumption

Chapter 2 discussed the challenges with innovating in sectors that have been born in captivity, or saddled with decades of stifling bureaucracy and regulations. What can be done to help those innovators break free of those constraints? One solution to freeing such technologies and sectors is an obscure existing provision of the Federal Communications Commission's code. Section 7 of the Communications Act (47 USC 157) was put in place in 1983 to encourage the more rapid adoption of "New Technologies and Services."

The first part of this FCC provision reverses the burden of proof regarding the need for regulation, and the second provision puts a timetable on action. The full language of the provision reads as follows:

(a) It shall be the policy of the United States to encourage the provision of new technologies and services to the public. Any person or party (other than the Commission) who opposes a new technology or service proposed to be permitted under this chapter shall have the burden to demonstrate that such proposal is inconsistent with the public interest.

(b) The Commission shall determine whether any new technology or service proposed in a petition or application is in the public interest within one year after such petition or application is filed. If the Commission initiates its own proceeding for a new technology or service, such proceeding shall be completed within 12 months after it is initiated.

My Mercatus Center colleague Brent Skorup refers to this provision as "The Innovator's Presumption," because it insists that entrepreneurs and their innovations be treated as innocent until proven guilty. The language of this provision could help many other emerging technologies break out of captivity.[39]

Section (a) of the provision puts the burden on opponents of change to make the case for preserving the regulatory status quo, which is important because many regulations are framed in such a way that nothing is allowed until some agency or official formally blesses that change. This FCC provision instead makes it clear that the opposite is the case and that the advocates of control must be able to make a compelling case that the benefits of intervention clearly outweigh those of ongoing experimentation. Until they can do so, permissionless innovation deserves the benefit of the doubt. In essence, innovation should be innocent until proven guilty. Section (b) of the provision is equally important because it imposes a stopwatch on the process.

The FCC provision's problem is that—although it puts the burden of proof on outside parties to make the case against freedom to innovate—it fails to impose the same requirement on the FCC itself. The presumption should place the same requirement on regulators. By default, they too should bear the burden of proving why innovations should be disallowed.

If that change occurred, this provision could serve as a model for how to make permissionless innovation a legal standard. To simplify matters and make the concept more widely applicable, we can generalize the Innovator's Presumption as follows: *Any person or party (including regulatory bodies) who opposes a new technology or service shall have the burden to demonstrate that such proposal is inconsistent with the public interest.*

Employ the Sunsetting Imperative to Periodically Clean House

There is no silver-bullet solution to the problems of regulatory accumulation and overcriminalization. The problems were decades in the making and will take many years to reverse. Mercatus Center economist Patrick McLaughlin has outlined several reforms that lawmakers could implement to begin tackling this serious problem, including legislative impact accounting, regulatory budgeting, regulatory review commissions, and hard caps on regulatory growth.[40] The Trump administration has undertaken some important regulatory reforms, including a "two-for-one" requirement requiring the elimination of two old mandates for each new one added.[41] But much more needs to be done.

Sunsetting requirements would be the optimal solution to this problem. "Congress must clean out the regulatory stables," Philip Howard correctly concludes. "Society can't function when stuck in a heap of accumulated mandates of past generations."[42] Periodic sunsets of old laws and regulations are particularly important in a world where modern technology markets are evolving at the speed of Moore's Law.

If innovative companies are expected to reinvent their business models every couple of years, it is reasonable to demand that the public policies governing those companies' technologies adapt at a similar pace. In essence, we need a Moore's Law for technology policy, or what we might think of as a Sunsetting Imperative. It would demand the following: *Any existing or newly imposed technology regulation should include a provision sunsetting the law or regulation within two years.*[43]

Policymakers can always reenact the rule or regulation if they believe it is still sensible. A similar provision should be put in place for older restrictions on existing technologies. This fresh-look requirement will help stem the tide of regulatory accumulation and ensure that only those policies that serve a pressing need remain on the books. Law professor Sofia Ranchordás argues as follows:

> Regulators can increase flexibility of regulations to accompany the pace of innovation both by including a sunset clause—which predetermines their expiry at the end of a certain period—or by experimenting with new rules. . . .

Terminating regulations by employing sunset clauses or by experimenting on a small-scale can be useful to ensure that rules keep up with the changes in technology and society.[44]

Coupled with the Innovator's Presumption advocated earlier, the sort of periodic housecleaning required by the Sunsetting Imperative will help ensure that technology policy doesn't shackle opportunities for entrepreneurial experimentation and economic growth.

The experience with sunsetting and de-licensing requirements at the state level has been mixed because, unsurprisingly, some legislatures have ignored or circumvented requirements.[45] In some cases, sunsetting rules have been effective in closing or curtailing some archaic laws or programs, but it will always be an uphill battle to ensure these rules are enforced.[46] My Mercatus Center colleague James Broughel has noted that some efforts to reduce red tape and regulatory accumulation, such as Virginia's Regulatory Reduction Pilot Program, have gained bipartisan support.[47] The Virginia measure aims to reduce regulations and other similar requirements by 25 percent over a three-year period, primarily by examining excessive occupational licensing rules and other examples of overcriminalization. If nothing else, the pressure of suggesting that sunsetting requirements be adopted can help keep some policies and programs in check.

Use the Parity Provision and Reciprocity Agreements to Level the Playing Field

Asymmetrical regulation is a major problem in many sectors, and it holds back entrepreneurialism and many important innovations. The problem arises whenever an entrepreneur comes up with a disruptive new concept and invades an old sector.

Level-playing-field problems are inevitable because the defenders of traditional regulatory regimes—which can include not only policymakers but also industry incumbents—have an interest in propping up long-entrenched bureaucracies and permitting schemes. Those interests will argue that when disruptive innovators invade their turf, it represents unfair competition because innovators might not face the same regulatory constraints.

Writing about this problem in the context of financial technology markets, my colleague Brian Knight has shown how "[i]ncongruous regulation could place new entrants at an undue disadvantage compared to their incumbent competitors and may deprive consumers of a fully competitive market."[48] That view applies equally to every other sector and technology, especially wherever a crazy-quilt of state and local licensing rules encumber entrepreneurial efforts. Technology trade law expert Anupam Chander refers to this situation as the problem of "technological neutrality."[49]

Calls to level the playing field or achieve neutrality in the direction of more red tape for the new players are misguided, however. Two wrongs do not make a right, and there is no reason we need to double down on the inefficient and ineffective regulatory regimes of the past in the name of leveling the proverbial playing field.

The better way to solve this problem and work toward parity is to borrow a page from trade law by adopting the equivalent of a most-favored-nation (MFN) clause for technology policy. MFN status has been crucial to the success of free trade agreements for many decades because it demands that nations accord equal, nondiscriminatory treatment to similar goods imported from other countries that are part of such treaties.

Importantly, this equal treatment is achieved by harmonizing in the direction of greater trade freedom, not more restrictions.

The World Trade Organization defines MFN treatment as follows: "If a country improves the benefits that it gives to one trading partner, it has to give the same 'best' treatment to all the other WTO members."[50] In other words, it levels the trade playing field by demanding reciprocal, nondiscriminatory treatment of all players included in a treaty. In recent years, that policy has come to be more commonly known as "permanent normal trade relations."

A similar MFN or normal trade relations clause could be adopted for technology policy to remedy the problems created by regulatory asymmetries. Such a regulatory "Parity Principle" would state that *any operator offering a similarly situated product or service should be regulated no more stringently than its least regulated competitor.*

What this principle means in practice is that policymakers should look to level the playing field by "deregulating down" to put everyone on equal footing, not by "regulating up"; this principle will achieve the same goal by a better means. They should attempt to relax old rules on incumbents as new entrants and new technologies challenge the status quo while also ensuring that new entrants face only minimal regulatory requirements as rules are reformed.[51]

Reciprocity provisions can help spur new entry and competition in other ways. As already noted, it would be helpful to have reciprocity among the states for occupational licensing requirements. Licensed workers need new permits in each jurisdiction to engage in many undertakings. That limits worker mobility, economic opportunity, and the right to earn a living.[52]

Although it will often be difficult to achieve perfect regulatory parity in practice, the aspirational goal of placing all players and technologies on the same liberalized level playing field should be at the heart of technology policy.

Some critics will claim that achieving parity in that way could lead to a "race to the bottom" in certain instances by disallowing state action and the application of local laws and norms. But deregulatory parity is not synonymous with anarchy. As the experience with the sharing economy illustrates, laws can be reformed with an eye toward achieving parity while simultaneously advancing various political objectives. The key is to do so in a way that enhances consumer welfare by expanding choice, competition, and the potential for permissionless innovation.

Use Litigation and an "Innovator's Defense Fund" to Defend Evasive Entrepreneurs

With so many vested interests and misguided public policies stacked against them, entrepreneurs need all the help they can get. Fortunately, many public-interest legal defense organizations exist that push back against the permission society by litigating against traditional barriers to innovation. For example, the Institute for Justice, the Goldwater Institute, and the Pacific Legal Foundation all routinely mount legal challenges to inefficient and unjust licensing, permitting, laws, and regulations. Yet a more focused legal defense effort may be needed as evasive entrepreneurialism and technological civil disobedience grow.

In his 2016 book, *By the People: Rebuilding Liberty without Permission*, Charles Murray proposed "a program of systematic civil disobedience underwritten by privately funded legal resistance to the regulatory state."[53] Building on that notion, in 2016, Eli Dourado and I proposed the creation of an Innovator's Defense Fund, a nonprofit law firm or legal fund tasked with defending innovators whose entrepreneurial efforts are stifled by the requirement to get government permission.[54] The idea would be to build a base of support for future innovators who lack the resources to defend themselves when they work around outdated and inefficient regulations. Optimally, such a legal defense fund would be backed by philanthropists and foundations who were devoted to using legal activism to defend the freedom to innovate.

An Innovator's Defense Fund could take two different forms. A full-service advocacy organization would be the most ambitious approach. Under this model, the organization would include many different legal specialists and have the ability to address a wide variety of issues involving evasive entrepreneurs operating in many different sectors. A less ambitious approach could take the form of a smaller organization with a board of directors and a handful of staff who identified clients and connected them with specific legal advocates who could address the nuanced facts and technical issues associated with each case.

Other organizations and educational institutions are already developing legal defense strategies for entrepreneurs. Launched in 2018, Trust Ventures is a venture capital firm that aims to "find, fund, and assist companies whose groundbreaking products, services, and innovations would otherwise be locked out of the marketplace by burdensome public policy barriers."[55] It is led by Salen Churi, who also directed the Innovation Clinic at the University of Chicago law school. The Innovation Clinic brings together law students and startup companies to devise legal strategies for navigating complex business and policy issues.[56] Similar clinics exist at other law schools. For example, the Santa Clara University law school created an Entrepreneur's Law Clinic and the University of Minnesota law school offers an Intellectual Property and Entrepreneurship Clinic.

Business schools have created similar programs to assist entrepreneurs. For example, the University of Toronto Creative

Destruction Lab was formed in 2012 at the Rotman School of Management as "a seed-stage program for massively scalable, science-based companies."[57] The lab has now expanded to include locations in five other cities in conjunction with other universities, including New York University's Stern School of Business. The program helps entrepreneurs find investors and get assistance from top MBA students.

Other innovation advocacy efforts are underway. Earlier in the book, we met Bradley Tusk, who helped Uber craft a strategy to overcome barriers to the New York City taxi market. In 2015, Tusk founded Tusk Ventures, which bills itself as "one-part venture capital firm, one-part political strategy firm that deploys both capital and political expertise to help startups break through regulatory barriers and operate at full speed."[58] The firm notes that "politicians and entrenched interests typically don't thank you for the disruption, they punch back," and it says its mission is to "protect startups from politics."[59] Finally, the Institute for Justice has also published an *Entrepreneur's Survival Guide* outlining the steps innovators can take to defend their rights in court and legislatures.[60]

Although much more needs to be done, these programs and initiatives make it clear that help is out there for innovators looking to disrupt the status quo.[61] Entrepreneurs should defend their rights more aggressively when the freedom to innovate and earn a living is being infringed.

Respect Consumer Choice, the Right to Try, and the Freedom to Tinker

The previous recommendations focused on creating more innovation opportunities for innovators and workers. It is equally important to ensure that citizens have the right to enjoy new goods and services and to use them as they choose.

People often speak of having a right to specific goods and services, but they rarely speak of the right to the process of innovation more generally. Within just the past quarter century, the internet went from being something that was feared or mocked as unnecessary, to an interesting novelty, to something that we now want everyone on the planet to be able to access cheaply or for free. But before we can demand that any technology become an entitlement, it first has to be invented, and the key to invention is incessant trial-and-error experimentation. This is why the freedom to innovate is so important for all of us. We must protect the process, not just the results, of innovation.

The freedom to innovate is also important in the downstream direction because of how average people use new goods and services. Once life-enriching and life-saving technologies are invented, the public often quickly comes to adopt them, tinker with them, and use them in creative combination with other technologies. Consider the myriad ways the public has used computers, broadband connections, websites, smartphones, social networking platforms, digital cameras, and hosting services to create a staggering array of new services and forms of culture. Also consider the many interesting ways that the public has tinkered with each of these technologies to improve their capabilities or tailor them to their needs.

The open-source movement pioneered many of the most exciting digital applications of our time largely through a can-do spirit of bottom-up, community-based knowledge sharing. A collaborative, crowd-sourced online encyclopedia like Wikipedia would have sounded outlandish to most people in the 1970s, when massive, multivolume physical books produced by professionals seemed like the only game in town. Today, however, Wikipedia is often the first stop in our search for information about the staggering array of topics we can research online.

We need to reframe and broaden our concept of the freedom to innovate to incorporate the freedom for all people to experiment with new and old technologies alike. Creativity is fueled by inquisitive minds, serendipitous interactions, and the incessant need for humans to strive to do more and better things for ourselves, our loved ones, and the world. We will go on constantly surprising ourselves and each other if we unleash the innovative spirit of all people. Julian Simon and Herman Kahn, two of the great scholars of economic progress, explained this principle best. "The ultimate resource is people—skilled, spirited, and hopeful people who will exert their wills and imaginations for their own benefit, and inevitably they will benefit not only themselves but the rest of us as well," argued Simon.[62] Once they do, wonderful, serendipitous, completely unpredictable things happen, as Kahn and a coauthor explained more than a half century ago:

> History is likely to write scenarios that most observers would find implausible not only prospectively but sometimes, even in retrospect. Many sequences of events seem plausible now only because they have actually occurred; a man who knew no history might not believe any. Future events may not be drawn from the restricted list of those we have learned are possible; we should expect to go on being surprised.[63]

Kahn's assertion that someone who "knew no history might not believe any" is an apt description of the Wikipedia phenomenon. Similarly, who would have believed that a dispersed community of coders could cobble together sophisticated operating systems for computers and smartphones? And yet Linux runs on virtually all supercomputers and is ubiquitous in the architecture of the internet. Few would have believed such a thing was possible unless they had enjoyed the fruits of those creative endeavors themselves and seen how well they work (and, more important, how quickly they are improved).

Building on the general "freedom to innovate," our expanded rallying cry should also incorporate the "right to try" and the "right to tinker." In some cases, these "rights" can be bolstered with a legal defense, and some formal "right to tinker" laws have been proposed from time to time. Most are somewhat context specific and involve questions about whether consumers should be able to tinker with or repair their own devices.[64] But it would be useful to broaden these rights and make them more generally applicable for consumers in other contexts.

Embrace Entrepreneurial Administration and Multistakeholder Processes

Another theme I have stressed throughout the book is how important it is for governments to avoid overzealous responses to evasive entrepreneurialism. Wishful thinking about enforcing outdated policies at any cost is unwise and will likely only result in a perpetuation of the compliance paradox, in which expanded enforcement efforts result in more legal evasion.[65] Even when crackdown efforts bring evasive entrepreneurs to heel, crackdowns can backfire in other ways. They could limit the sort of new business formation and competition that can help expand innovation opportunities, jobs, and economic growth. Most policymakers do not want that result, and so they will need to think about different responses and approaches to evasive activities. Meanwhile, the pacing problem will continue to rear its head with increasing regularity. With innovators "moving at the speed of imagination," as former FAA Administrator Michael Huerta has noted, regulators "can't afford to move at the old speed of government. We have to be willing to innovate the way we do our work," he correctly observed.[66]

That need is why this book has stressed the importance of fresh thinking about technological governance. "Entrepreneurial administration" may sound like just another catchphrase that is doomed to fail, but it can have real significance. "Regulators must become as creative as innovators," argues regulatory expert Richard Williams.[67] In the abstract, thinking entrepreneurially about government demands a willingness by policymakers to be humble, patient, and open to change.

The specific elements of entrepreneurial administration and flexible regulation were already discussed at length in Chapters 6 and 7. But soft-law and multistakeholder processes in particular will be the linchpins of the new governance models for emerging technologies. As Ryan Hagemann concludes, "the advantages of governing emerging technologies using soft criteria and multi-stakeholderism seem to outweigh the potential costs. Soft law and its instruments of implementation have been a driving force behind the post-Internet approach to regulating new technologies and will likely continue to dominate the process by which such rules are developed," he says.[68] This tendency is the case because, much like how "the traditional command-and-control approach was viewed as an untenable governance mechanism for the early internet,"[69] that approach has also become unworkable and unwise for most other emerging technologies.

Innovation advocates should be willing to work with lawmakers and regulators to devise more flexible, adaptive, cost-effective strategies for addressing pressing policy issues. Advocates should also be willing to engage in such negotiations even when they might be working simultaneously to undo inefficient or outdated regulatory regimes altogether. Going for broke with all-or-nothing regulatory reform strategies, especially agency abolition proposals, is unlikely to bear fruit, and backup reform strategies like collaboration with lawmakers would be wise.

9

CONCLUSION:
THE CASE FOR RATIONAL OPTIMISM

Joe Carlen concluded his *Brief History of Entrepreneurship* with the prediction that "the transformational impact of entrepreneurship will persist—as undeniably profound and inescapably controversial as ever."[1] The continued expansion and combination of new technological capabilities and different types of evasive entrepreneurialism will make Carlen's prediction an absolute certainty. As this process unfolds and opposition to technological change intensifies, the freedom to innovate will need a stronger defense than ever before.

This book has attempted to offer that defense. I have argued that technological innovation is immensely valuable to society and, by extension, innovative activities—even those of an evasive nature—deserve more appreciation and support. The act of being creative, seeking out new opportunities, and finding new and better ways of doing things—that is, being entrepreneurial—is valuable because it improves the welfare of the people, their societies, and even their governments. The technologies discussed throughout this book have the potential to massively improve human well-being by giving us access to tools and capabilities that are life enriching and life extending. These innovations can also bring down the cost of living for all citizens.[2] Beyond providing society with goods and services that help us improve our living standards over time, technological innovation also helps people live lives of their own choosing and earn a living as they see fit. People then have the means to pursue happiness and meaning—however they define it.

If, in the process, entrepreneurialism and innovation can help us break political logjams and force governments to become more adaptive and accountable, then all the better. A certain amount of evasive entrepreneurship, I have argued, can act as a sort of relief valve or circuit breaker to counteract negative pressures in the system before things break down completely. By challenging legislators and regulators to reevaluate the wisdom of their policies, evasive entrepreneurship and technologically enabled civil disobedience can actually help "spur institutional change with potentially important welfare effects."[3] It can help bring law in line with new realities and make sure it more accurately reflects the consent of the governed.

For all those reasons, I have argued that there is a powerful *moral* case in favor of the general freedom to innovate. I have also conceded that it is wise to think about the challenges and dangers posed by new technologies, but I have argued that we should do so with humility, patience, and flexibility in mind. While looking to address some of the tradeoffs posed by technological innovation, we must not fall prey to apocalyptic rhetoric, dystopian narratives, and technopanic thinking. If we base innovation policy on the idea that every hypothetical worst-case scenario must be addressed before we move forward, then many *best-case* scenarios will never come about.[4]

A Positive Vision: Rational Optimism

Perhaps the best label for the vision I have laid out here is "pragmatic optimism," or "rational optimism" to borrow Matt Ridley's preferred term for this worldview.[5] As I suggested in my previous book, rational optimists have the following in common:

- They believe that innovation, economic growth, pluralism, and human betterment have a symbiotic relationship but are not naive about the challenges sometimes associated with technological change.
- They look forward to a better future and reject "rosy retrospection bias" (i.e., emotional appeals to slow progress on the basis of nostalgic accounts of some supposed "good ol' days" or bygone eras).
- They base this optimistic outlook on empiricism and historical analysis, not on blind faith in any particular viewpoint, ideology, or gut feeling.
- They advocate practical, bottom-up solutions to hard problems through ongoing trial-and-error experimentation but aren't wedded to any one process to get the job done.
- They greatly appreciate entrepreneurs' willingness to take risks and try new things but don't engage in hero worship of any particular individual, organization, or technology.

Rational optimism is a pro-growth, pro-progress, pro-freedom vision of cultural and economic change, but it is not Pollyannaism, technophilia, or utopianism. In fact, rational optimists are *anti*-utopians, because they understand that hard problems can only be solved by better appreciating and learning from the error part of trial and error. No one is divinely inspired to come up with the One Best Way of doing things from the beginning.[6] Wisdom, as well as progress, is a function of our willingness to tolerate and learn from our failures. Writing in 1960, Nobel Prize–winning economist F. A. Hayek wisely observed that many intellectuals "ignore the importance of the freedom of *doing* things" and that "[f]reedom of action, even in humble things, is as important as freedom of thought."[7] Indeed, the two are inextricably linked. Learning by doing might more accurately be called *learning by failing.*[8]

Alas, too many technology critics tend to disparage this learning process or draw the wrong lessons from it. To the extent that critics are interested in entrepreneurialism, experimentation, and the many failures involved, it is often to snobbishly deride that process. They may suggest that every failure is an example of why we can't trust that process at all.[9] In reality, ongoing experimentation is the key to unlocking knowledge and prosperity—for individuals, organizations, nations, and entire civilizations.[10] We need lots and lots of experiments—and many corresponding failures—to figure out what works best.[11] Learning from our experiences and failures will ultimately help us to live wealthier, healthier, safer, and longer lives.[12]

"We Need to Have a Conversation"

Technology critics will still protest and suggest that rational optimists fail to take into account many other important issues, social values, or human rights. "We need to have a conversation," they will insist. Almost every article or book written by tech critics today includes some sort of demand that we "have a conversation" about new technologies before they are allowed to go forward.[13] As part of those "conversations," the critics want us to ask and answer questions like "what kind of technologies do we really want?"[14] or "what is the end" that any given technology serves or will bring about for society?[15]

By all means, let's have those conversations. But let us also be realistic about a few things when we do.[16] Such inquiries typically assume that we possess adequate knowledge to understand or forecast future technological trends and social needs, which is almost never the case. Karl Popper, the great philosopher of science and politics, once observed that the problem with what he called "the Utopian programme" of planning for the future is that "we do not possess the experimental knowledge needed for such an undertaking."[17] Hayek devoted much of his life to proving the same point, noting in his famous essay "The Use of Knowledge in Society" that the practical problem with proposals to plan progress in a top-down fashion is that "the knowledge of the circumstances of which we must make use never exists in concentrated or integrated form but solely as the dispersed bits of incomplete and frequently contradictory knowledge which all the separate individuals possess."[18]

Thus, a conversation about the future of any new technology or technological process is bound to run up against this "knowledge problem," limiting our ability to make wise choices about future uncertainties. Again, trial-and-error is important; experiments and experience give us answers that forecasting cannot.

The other obvious problem with critics insisting that we should have a conversation before allowing innovations to move forward is that it is also unclear *who* the "we" is in all these conversations and *when* and *where* "we" are having all these conversations. Finally, and perhaps most problematically, it is not at all clear when such conversations are adequately concluded.

Consider this thought experiment. The year is 1888, and some concerned critics begin asking the question, "What kind of technologies do we really want?" A conversation ensues (presumably first in universities, then in major books and papers, and then later in legislative bodies) that identifies several things that "we" deem desirable or not. Perhaps some existing or anticipated agricultural and industrial technologies are debated. It is unlikely, however, that anyone asks about a portable camera and photography. These things are about to explode onto the scene, however, as George Eastman is inventing the Kodak roll-film camera in that year.

When a conversation *did* ensue about cameras a few years later, it was a decidedly negative one—at least among elites.[19] Many of them wanted public photography regulated in some fashion (even though that would have raised some serious free speech issues).[20] While some elites were wringing their hands about the camera and wondering what should be done about it, however, many others had decided that they were going to go ahead and buy cameras for personal use and make them part of the human experience. Cameras would quickly become a cherished part of people's lives, allowing them to capture, share, and retain important moments and scenes of the world around them. Obviously, there would still be problems with ubiquitous cameras (snooping, paparazzi, and other intrusions), but over time society devised sensible social and legal norms to address some of the worst-case problems while allowing the many best-case uses of photography to play out. In other words, society muddled through and moved on after a lot of trial and error with portable cameras.

Here's another example. If someone in the 1960s had asked, "What kind of technologies do we really want?," it is unlikely that there would have been much discussion about the internet, smartphones, 3D printing, or commercial drones. We were pre-occupied at the time with debates over analog-era technologies like television and radio. We were no better equipped to "have a conversation" about these modern technologies than we were when the portable camera came along. When we imagined the future, we imagined flying cars, jetpacks, and even living on other planets.

The point is not that conversations about the future of technology are worthless. Technological assessment and criticism, and the deliberations that they entail, can play a valuable role in helping us think through the challenges associated with technological change. We should not, however, expect too much practical wisdom or effect to come from such conversations—at least not too much preemptive wisdom. Real wisdom is born of real-world human experiences, endless trial-and-error, and the resiliency that goes along with muddling through adversity.[21]

Yale political scientist James C. Scott has written eloquently about the importance of what ancient Greeks referred to as *metis*, or the "wide array of practical skills and acquired intelligence in responding to a constantly changing natural and human environment."[22] Scott has documented how various cultures and communities have prospered in the face of adversity, thanks to a combination of practical skills, localized knowledge, folk wisdom, and shorthand rules of thumb.

Learning by doing through endless trial-and-error experimentation and improvisation allows individuals, organizations, governments, and entire societies to become more resilient over time. That way of operating goes back to the distinction stressed throughout this book about the benefits of bottom-up as compared to top-down approaches to governance. Table 9.1 attempts to summarize how those approaches differ in what each model or vision stresses.

TABLE 9.1
Summary of two approaches to governance

Top-down governance	Bottom-up governance
Promulgated rules	Practical rules of thumb
Anticipatory governance	Responsive governance
Build-and-freeze regulation	Adaptive, flexible rules
Centralized knowledge	Localized knowledge
Fragility of systems	Resiliency of systems
Stability and equilibrium	Spontaneity and experimentation
Monocentric order	Polycentric order

Scott's insights are particularly wise when we consider technological governance and the massive uncertainties it entails. We do not possess crystal balls that will allow us to forecast the technological future, especially in an age of "radical" or "seismic" innovations.[23] Technological forecasting is mostly guesswork based on past assumptions, and those assumptions are particularly unlikely to hold for fast-moving technology sectors where hyperuncertainty is the norm.[24] "The past seldom obliges by revealing to us when wildness will break out in the future," risk historian Peter L. Bernstein once noted.[25] Consequently, forecasting can have profound costs when done wrong.

More decentralized and experimental approaches can therefore help us achieve better results. Those successes are why scholars like Nobel laureate Elinor Ostrom and others have stressed the importance of "polycentricity" in the governance of complex systems.[26] Polycentricity represents "a social system of many decision centers having limited and autonomous prerogatives and operating under an overarching set of rules."[27] More simply, we might think of polycentricity as flexible governance. Flexible governance attitudes are particularly useful when in a conversation about the future of governance because—even when we successfully anticipate the development of a particular technology—we might not be able to agree on common metrics or acceptable baselines about how to evaluate potential benefits and costs.

For example, in the 1970s, many nations *did* have a robust conversation about nuclear power as an energy source. One could argue that some countries, such as the United States, made a serious miscalculation in that conversation by discouraging nuclear technologies, and that this mistake has made it harder to reduce global warming by replacing fossil fuels with cleaner nuclear power sources.[28] It's also true that many people involved in that conversation could not come to agreement about common metrics regarding the sensibility or safety of nuclear power. But the warming of the atmosphere was not the primary consideration at the time those discussions took place; it was mostly a conversation about the safety of nuclear power plants and nuclear waste disposal. Keep in mind, these were conversations about energy sources that were already well understood and operational. Moreover, energy markets and energy innovations were slower to come about as compared with most of the newer technologies discussed throughout this book. Despite that slower pace, we got things wrong in those conversations, or we at least failed to account for the full range of future issues that should have been part of those discussions.

In sum, it is important to realize that to the extent that informal *societal* conversations about technologies become *political* conversations with some presumed preemptive solutions, those conversations will be based on limited knowledge about complex issues and values, both in the present and, especially, in the future. At some point, therefore, we must move forward and adapt ourselves to new realities and use more flexible governance arrangements to do the best we can to guide technological developments.

Let the Adventure Continue

Rational optimists must make the case for technological change again and again because, as history has illustrated, tech critics are relentless in pushing for whatever their preferred status quo may be at any given point in time. Despite the overwhelming empirical evidence that innovation has driven remarkable improvements in human well-being, most critics and many average citizens, for whatever reason, either ignore that evidence or simply refuse to believe it.[29] Pessimism bias runs deep in our culture despite the cornucopia of riches that innovation has given us.

Therefore, rational optimists must redouble their efforts to defend the importance of innovation. They must document the dangers of technopanic thinking and public policies based on the precautionary principle. To reiterate, rational optimists are not calling for a rejection of all technological governance efforts. Rather, they are calling for a rejection of status quo thinking and the stasis mindset that would lock society into old, broken ways of doing things—both in markets and in government.

Calestous Juma concluded his masterwork *Innovation and Its Enemies* by reminding us of the continued importance of "oiling the wheels of novelty" to constantly replenish the well of important ideas and innovations. "The biggest risk that society faces by adopting approaches that suppress innovation," Juma said, "is that they amplify the activities of those who want to preserve the status quo by silencing those arguing for a more open future."[30] The openness Juma had in mind represents a tolerance of new ideas, inventions, and unknown futures.[31] It can and should also represent an openness to new, more flexible methods of technological governance.

Similarly, almost 40 years ago, the engineer Samuel C. Florman addressed the anti-technology movement of his time in his book *Blaming Technology: The Irrational Search for Scape-goats*. He concluded by noting the profound danger of giving up on finding new and better ways of doing things:

> By turning our backs on technological change, we would be expressing our satisfaction with current world levels of hunger, disease, and privation. Further, we must press ahead in the name of the human adventure. Without experimentation and change our existence would be a dull business.[32]

We must embrace the "open future" and "human adventure" that Juma and Florman described if we hope to continue to prosper as a species. We will only attain that goal by vociferously defending the freedom to innovate and the entrepreneurial spirit that propels the progress of our civilization.

As this book was heading to the printer, the coronavirus that causes COVID-19 spread across the globe with devastating effects. The pandemic caused worry and frustration everywhere, including in the United States. While the United States prides itself on being a land of innovation, when it comes to medical innovation, the unfortunate reality is that America sometimes lags other countries in important ways. Approval of medical devices and medical testing procedures are prime examples. As noted throughout this book, the U.S. system is remarkably precautionary in this regard. A complex web of Food and Drug Administration (FDA) regulations and Centers for Disease Control and Prevention (CDC) procedures encumber attempts to innovate rapidly or in unique ways. Risk-averse regulators and their build-and-freeze rules prioritize conformity to existing processes and frown upon challenges to the status quo.

When the coronavirus hit, the regulatory process broke down, and the problems associated with the old rules came into public view in dramatic fashion. Journalists and average citizens started asking why people in other countries were able to get tested so much more quickly and easily than in the United States. "Testing in the U.S. has been wholly inadequate," argued *Wall Street Journal* columnist Peggy Noonan. "History may come to see this as the great scandal of the epidemic."[1] Social media lit up with anger from citizens and experts alike about the political and regulatory failures they were witnessing.

After weeks of waiting for federal agencies to develop effective tests or greenlight private efforts, some people started searching for testing solutions or care options, even if regulators had not yet given them formal approval to do so. Once again, necessity became the mother of invention. Evasive entrepreneurs stepped in.

For example, in early March 2020, with fears about the coronavirus spreading domestically, Dr. Helen Y. Chu and a team of infectious disease experts in Seattle began wondering if they might have a novel way to test for the virus by repurposing a flu test they had already developed.[2] The only government-approved diagnostic tests were contaminated, meaning that federal officials should have welcomed credible alternatives.[3] When Chu's team offered a workable alternative procedure, however, federal bureaucrats were not interested in the help.[4]

A *New York Times* investigation into the episode revealed that "nearly everywhere Dr. Chu turned, officials repeatedly rejected the idea," even though it was clear that the virus was already ravaging China and Italy and was likely spreading rapidly in the United States. She and her team decided to start performing coronavirus tests without government approval. Their worst fears were confirmed when they were able to document that a local teenager had the virus. Government officials later confirmed the findings from Chu's team.

According to *Times* reporters, regulators would not allow Chu's team to repurpose its test because the test had not gone through traditional regulatory channels and because the lab lacked the proper certifications for clinical work.[5] Chu begged the regulators for testing flexibility; she received support from her backers at the Bill and Melinda Gates Foundation as well as health officials from the state of Washington. They all petitioned the CDC, asking the agency to consider using the Seattle team's test results. The CDC simply punted on the issue and told them to seek FDA approval. Because Chu's team did not strictly comply with established CDC and FDA regulations, however, federal regulators simply refused to budge. "We felt like we were sitting, waiting for the pandemic to emerge," Chu told the *Times*. "We could help. We couldn't do anything." Federal officials told them to stop testing altogether.

Times reporters Sheri Fink and Mike Baker noted that the government's failure to allow Chu's test to go forward "was just one in a series of missed chances by the federal government to ensure more widespread testing during the early days of the outbreak, when containment would have been easier."[6] The reporters said the episode "illustrates how existing regulations and red tape—sometimes designed to protect privacy and health—have impeded the rapid rollout of testing nationally, while other countries ramped up much earlier and faster."

Other scientists and organizations felt the same regulatory pinch as Chu. Dr. Melissa Miller, director of the Clinical Microbiology Laboratory at University of North Carolina Hospitals, told *NBC News*, "I could have tested over 1,000 patients by now instead of checking boxes."[7] At the time, Miller said her lab had developed a test based on the World Health Organization's protocol before the state of North Carolina's official lab even had its own. FDA paperwork requirements held back her innovative efforts, just as they did for Chu. "Faced with a public health emergency on a scale potentially not seen in a century," Fink and Baker conclude, "the United States has not responded nimbly."[8] Well-intentioned precautionary regulations had unfortunately prevented potentially helpful rapid responses to a growing public health emergency.

Chu's evasive act of testing without approval also tees up the core ethical and legal question raised repeatedly throughout this book: What are the ethics of evasive activities, and why is it that many people often justify acts of evasive entrepreneurialism after the fact but few defend them as they are happening? Why were many people (including *New York Times* journalists) generally supporting Chu and her team instead of suggesting they be fined or sent to jail? After all, technically speaking, they broke the law. Some will say Chu and her team can be forgiven because their evasive actions could have helped save lives, and therefore, the ends justified the means.

That is not the way the law works, however. Compliance comes first under the existing regulatory system, even if the system is holding back lifesaving progress. Some would still insist that they should be free from regulatory persecution or prosecution following their efforts to develop a successful test. But who would have stood by Chu and her team when they *initially* set out to innovate around the broken system to offer a credible alternative? This is why I have suggested throughout this book that evasive entrepreneurs deserve more support as they are challenging the status quo, not just well after the fact.

Incidentally, Chu and her team were not the only ones engaging in evasive efforts to work around red tape to serve the public good. Doctors at the Johns Hopkins Hospital in Baltimore were simultaneously working to create their own test for the coronavirus, for example.[9] As the crisis intensified, social media sites were also suddenly filled with discussions about how people might come together to build open source tools and information to assist with virus testing or treatments. People discussed how they might be able to build DIY home ventilators using off-the-shelf hardware and open source software.[10] In Italy, where the virus overran the health system very rapidly, some hospitals began running out of valves for resuscitation devices. Local 3D printing firms responded to a call for help and brought printers to local hospitals to replicate the needed components quickly and cheaply.[11] In Hong Kong, people used 3D printers to replicate filters for more sophisticated respiratory masks, and university labs used them to make transparent medical face shields for doctors and nurses to wear.[12] Other average citizens both here and abroad openly debated how various biohacking solutions might contribute knowledge about alternative virus treatments. Again, all these activities butted up against long-standing regulations governing medical devices and therapies.

Even something as seemingly mundane as making or traveling with hand sanitizer was suddenly open for reconsideration. Many Americans probably did not realize that even hand sanitizer was under the FDA's regulatory purview. It likely came as a surprise to many people when the agency announced it would not take any enforcement action against facilities or licensed professionals who make hand sanitizer for consumers, so long as they adhered to a recipe laid out by the agency.[13] Plenty of people were already circulating homemade hand sanitizer recipes on various websites. They were not going to wait for federal officials to give them permission to make something as simple as a cocktail of alcohol, glycerol, and hydrogen peroxide. At the same time, the Transportation Security Administration announced that it was waiving the four-ounce limit for liquids and gels for hand sanitizer on airplanes. Critics pointed out that the rule change called into question the wisdom of the agency's 14-year ban on such small amounts of liquids.[14]

Some state politicians were simultaneously responding with flexible reforms that cut through red tape that limited the licensing of new medical personnel. For example, some states like Massachusetts looked to loosen occupational licensing laws to allow out-of-state medical professionals and nurses to receive license approval in just one day.[15] That was a positive step meant to improve health outcomes, but it raised the question: What were the negative effects of overly restrictive occupational licensing reforms *before* that reform? Regardless, the move by some governments to engage in such rule departures in the wake of the pandemic was consistent with the themes discussed throughout this book: sometimes even lawmakers and regulators will ignore their own rules when they no longer make sense or stand in the way of progress.

What matters about these various ideas and examples is not that any one of them was going to offer a miracle cure for the dangerous virus—they weren't. Effective response to global pandemics requires multifaceted solutions and considerable public and private resources. Nonetheless, the evasive entrepreneurialism and technological civil disobedience witnessed in the pandemic were important because they put positive pressure on a broken system.

There will always be a need for laboratory guidelines and medical device regulations, of course. Many rules make sense because there are serious risks associated with medical device experimentation or alternative therapies. But what happens when existing rules become so ossified and archaic that sensible types of experimentation are only allowed during a crisis? It should not take a public health emergency to freshen old rules and reconsider broken regimes. Regardless, when enough people come to realize a regulatory system undercuts their ability to live happy, healthy, and prosperous lives, they will increasingly use whatever new technological capabilities are at their disposal to evade misguided, inefficient rules.

In the case of Chu and her team's workable testing alternative, highly restrictive procedures for virus testing had the unintended consequence of shutting down tests that could detect outbreaks and save lives. For federal officials, going by the book apparently mattered more than getting good results. "This virus is faster than the FDA," a University of Washington Medical Center doctor told the *Times*. That doctor had petitioned the FDA to help Chu get her team's test approved rapidly. But regulators held the line. At one point, the agency even asked the doctor to submit materials through the mail, which would have added days or weeks to getting an answer. Strict regulatory paperwork procedures had triumphed over common sense.

That mentality needs to end. Policymakers need a new approach for technological governance that is more in line with modern realities. Flexibility and humility will be essential. Regulators do not need to throw out the old rulebooks altogether. Some precautionary procedures still make sense. But why not embrace the entrepreneurial spirit of the citizenry and allow more experimental trials, flexible testing procedures, and perhaps even prizes for particularly innovative ideas? When enforcing the rules that remain on the books, policymakers should also consider targeted waivers and ex post–regulatory reviews as opposed to ex ante–regulatory prohibitions on any and all evasive innovations. Liability rules can also be tweaked so that innovators do not have to live in constant fear of getting sued for trying to make the world a better place. Finally, post-market monitoring and recall notices can also be used to ensure flexible experiments have some regulatory guardrails.

But shutting down creative solutions and unique thinking because they run counter to some crusty old rulebook is never the right response. We should view evasive entrepreneurialism as an important part of a broader discovery process that incorporates the profound importance of ongoing, decentralized trial-and-error experimentation in the process of societal learning and improvement. Lawmakers should find a way to accommodate a little more outside-the-box thinking and innovating, and not just when our lives are on the line.

I wish to thank the following individuals for the helpful feedback or research assistance they provided while I prepared this book: James Broughel, Daniel Castro, Neil Chilson, Arthur M. Diamond, Jr., Mohamad Elbarasse, Ryan Hagemann, Anne Hobson, Lynne Kiesling, Michael Kotrous, Gary Marchant, Eileen Norcross, Andrea O'Sullivan, Jordan Reimschisel, Will Rinehart, Marc Scribner, Jennifer Huddleston, Maurice McTigue, Trace Mitchell, Walter Valdivia, Caleb Watney, and several anonymous reviewers.

Introduction

1. Adam D. Thierer, "Embracing a Culture of Permissionless Innovation," Cato Institute, Cato Online Forum, November 17, 2014, https://www.cato .org/publications/cato-online-forum/embracing-culture-permissionless -innovation.

2. Timothy Sandefur, *The Permission Society: How the Ruling Class Turns Our Freedoms into Privileges and What We Can Do About It* (New York: Encounter Books, 2016).

3. Adam Thierer, *Permissionless Innovation: The Continuing Case for Comprehensive Technological Freedom*, 2nd ed. (Arlington, VA: Mercatus Center at George Mason University, 2016), pp. 8–9.

4. Eli Dourado, "Technologies of Control and Resistance: Making Sense of Our Stagnant Dynamism," *Medium*, November 4, 2011.

5. Albert H. Segars, "Seven Technologies Remaking the World," *MIT Sloan Management Review*, March 9, 2018.

6. Richard Adler, *Navigating Continual Disruption* (Washington: Aspen Institute, 2015), p. 1.

7. World Economic Forum, *Deep Shift: Technology Tipping Points and Societal Impact* (Geneva, Switz.: September 2015), p. 3.

8. Calestous Juma, *Innovation and Its Enemies: Why People Resist New Technologies* (New York: Oxford University Press, 2016); Elizabeth Pollman, "Corporate Disobedience," *Duke Law Journal* 68 (2019): 731.

9. Deirdre N. McCloskey, *Bourgeois Equality: How Ideas, Not Capital or Institutions, Enriched the World* (Chicago: University of Chicago Press, 2016), p. xxxiv.

10. Adam Thierer, "Muddling Through: How We Learn to Cope with Technological Change," *Medium*, June 30, 2014.

11. Vinod Khosla, "We Need Large Innovations," *Medium*, January 1, 2018.

12. Tyler Cowen, *Stubborn Attachments: A Vision for a Society of Free, Prosperous, and Responsible Individuals* (San Francisco: Stripe Press, 2018), p. 92.

13. Sander Wennekers and Roy Thurik, "Linking Entrepreneurship and Economic Growth," *Small Business Economics* 13, no. 1 (February 1999): 51.

14. Benjamin M. Friedman, "The Moral Consequences of Economic Growth," *Society* 43, no. 2 (January/February 2006): 20.

15. Tyler Cowen, "Is Innovation Over? The Case Against Pessimism," *Foreign Affairs* 95, no. 2 (March/April 2016), https://www.foreignaffairs.com/reviews/review-essay/2016-02-15/innovation-over.

16. Donald J. Boudreaux, "What's Your Moonshot?," Mercatus Center at George Mason University, *Mercatus Original Video*, November 16, 2017.

17. Art Carden and Deirdre N. McCloskey, "The Bourgeois Deal: Leave Me Alone, and I'll Make You Rich," in *Demographics and Entrepreneurship: Mitigating the Effects of an Aging Population*, ed. Steven Globerman and Jason Clemons (Canada: Fraser Institute, 2018), p. 449.

18. Timothy Sandefur, *The Right to Earn a Living: Economic Freedom and the Law* (Washington: Cato Institute, 2010).

19. Arthur C. Brooks, "The Secret to Human Happiness Is Earned Success," *Washington Examiner*, July 14, 2010.

20. Adam Thierer, "The Right to Pursue Happiness, Earn a Living, and Innovate," Mercatus Center at George Mason University, *The Bridge*, September 20, 2018.

21. John Blevins, "License to Uber: Using Administrative Law to Fix Occupational Licensing," *UCLA Law Review* 64 (2017): 847.

22. Clyde Wayne Crews Jr., "What Key Factors Influence Entrepreneurship and Business Growth? Ideas for Researchers," *Forbes*, June 20, 2018.

23. Charlan Nemeth, *In Defense of Troublemakers: The Power of Dissent in Life and Bu*siness (New York: Basic Books, 2018).

24. Gary E. Marchant, "The Growing Gap Between Emerging Technologies and the Law," in *The Growing Gap Between Emerging Technologies and Legal-Ethical Oversight: The Pacing Problem*, ed. Gary E. Marchant, Braden R. Allenby, and Joseph R. Herkert (Dordrecht, Neth.: Springer, 2011), p. 19. See also Juma, *Innovation and Its Enemies*, p. 14.

25. Adam Thierer, "The Pacing Problem and the Future of Technology Regulation," Mercatus Center at George Mason University, *The Bridge*, August 9, 2018.

26. Hal R. Varian, "Computer Mediated Transactions," *American Economic Review* 100, no. 2 (May 2010).

27. David Rejeski, "Public Policy on the Technological Frontier," in Marchant, Allenby, and Herkert, *The Growing Gap*, p. 57.

28. Even advocates of increased regulation acknowledge this fact. See Cristie Ford, *Innovation and the State: Finance, Regulation, and Justice* (Cambridge: Cambridge University Press, 2017), pp. 19–20.

29. Emanuela Carbonara, Francesco Parisi, and Georg von Wangenheim, "Legal Innovation and the Compliance Paradox," *Minnesota Journal of Law, Science & Technology* 9, no. 2 (2008): 842.

30. Philip J. Weiser, "Entrepreneurial Administration," *Boston University Law Review* 97 (2017).

31. Ryan Hagemann, "New Rules for New Frontiers: Regulating Emerging Technologies in an Era of Soft Law," *Washburn Law Journal* 57, no. 2 (Spring 2018): 244.

32. Timothy C. May, "The Crypto Anarchist Manifesto," November 22, 1992, https://www.activism.net/cypherpunk/crypto-anarchy.html.

33. Samuel C. Florman, *The Existential Pleasures of Engineering*, 2nd ed. (New York: St. Martin's Griffin, 1994), p. 57.

34. Those seeking such a treatment of the issues presented here might consider reading James Dale Davidson and William Rees-Mogg, *The Sovereign Individual: Mastering the Transition to the Information Age* (New York: Touchstone, 1999).

35. Jonathan Turley, "The Rise of the Fourth Branch of Government," *Washington Post*, May 24, 2013 ("The fourth branch now has a larger practical impact on the lives of citizens than all the other branches combined.").

36. Albert O. Hirschman, *Exit, Voice, and Loyalty* (Cambridge, MA: Harvard University Press, 1970), p. 117.

37. Thomas Jefferson, Letter to William Stephens Smith, November 13, 1787, in *The Works of Thomas Jefferson*, Federal Edition, vol. 5, ed. Paul Leicester Ford (New York and London: G. P. Putnam's Sons, 1904–05), viewed on Online Library of Liberty.

38. Arthur M. Diamond Jr., *Openness to Creative Destruction: Sustaining Innovative Dynamism* (Oxford: Oxford University Press, 2019).

Chapter 1

1. Calestous Juma, *Innovation and Its Enemies: Why People Resist New Technologies* (New York: Oxford University Press, 2016), p. 280. See also Kevin Kelly, *The Inevitable: Understanding the 12 Technological Forces That Will Shape Our Future* (New York: Viking, 2016), p. 15.

2. Cristie Ford, *Innovation and the State: Finance, Regulation, and Justice* (Cambridge: Cambridge University Press, 2017), p. 209.

3. Edward W. Lawless, *Technology and Social Shock* (New Brunswick, NJ: Rutgers University Press, 1977), p. 4.

4. Tyler Cowen, "The Genesis of the Tech Industry, and Vice Versa," *Bloomberg Opinion*, March 5, 2019.

5. Evgeny Morozov, "The Taming of Tech Criticism," *The Baffler* 27, March 2015.

6. Barry Schwartz, *The Paradox of Choice: Why More Is Less* (New York: Ecco, 2004).

7. Evgeny Morozov, *To Save Everything, Click Here: The Folly of Technological Solutionism* (New York: Public Affairs, 2013). See also Evan Selinger, "Too Much Magic, Too Little Social Friction: Why Objects Shouldn't Be Enchanted," *Los Angeles Review of Books*, January 8, 2015.

8. For a critique of such thinking, see Samuel C. Florman, *The Existential Pleasures of Engineering*, 2nd ed. (New York: St. Martin's Griffin, 1994), p. 16, and Marian L. Tupy, "Are We Suffering from a Crisis of Meaning?," HumanProgress.org, March 30, 2018.

9. Morozov, "The Taming of Tech Criticism."

10. Evgeny Morozov, "Stunt the Growth," *Slate*, January 22, 2015.

11. Giacomo D'Alisa, Federico Demaria, and Giorgos Kallis, eds., *Degrowth: A Vocabulary for a New Era* (London: Routledge, 2015).

12. Brett Frischmann, "There's Nothing Wrong with Being a Luddite," *Scientific American*, September 20, 2018.

13. David Auerbach, "It's OK to Be a Luddite," *Slate*, September 9, 2015.

14. Martin Ford, *Rise of the Robots: Technology and the Threat of a Jobless Future* (New York: Basic Books, 2015); Rory Cellan-Jones, "Robots on the March," *BBC News*, July 2, 2015.

15. Florman, *The Existential Pleasures*, pp. 49, 70.

16. Franklin Foer, *World without Mind: The Existential Threat of Big Tech* (New York: Penguin Press, 2017).

17. Marc Goodman, *Future Crimes* (New York: Doubleday, 2015), p. 392.

18. Foer, *World without Mind*.

19. Wendell Wallach, *A Dangerous Master: How to Keep Technology from Slipping beyond Our Control* (New York: Basic Books, 2015).

20. Wallach, *A Dangerous Master*. See also David Brooks, "Our Machine Masters," *New York Times*, October 30, 2014.

21. Ben Austen, "The Terminator Scenario: Are We Giving Our Machines Too Much Power?," *Popular Science,* January 13, 2011; John Markoff and Claire Cain Miller, "As Robotics Advances, Worries of Killer Robots Rise," *New York Times*, June 16, 2014.

22. Shoshana Zuboff, *The Age of Surveillance Capitalism: The Fight for a Human Future at the New Frontier of Power* (New York: Public Affairs, 2019).

23. L. M. Sacasas, "The World Will Be Our Skinner Box," *Technology, Culture, and Ethics* (blog), November 19, 2018.

24. Brett Frischmann and Evan Selinger, *Re-Engineering Humanity* (Cambridge: Cambridge University Press, 2018), p. 250.

25. Alvin Toffler, *Future Shock* (New York: Random House, 1970), p. 2.

26. Toffler, *Future Shock*.

27. David Shenk, *Data Smog: Surviving the Information Glut* (San Francisco: Harper, 1997).

28. Richard Saul Wurman, *Information Anxiety 2* (Indianapolis, IN: Que, 2001).

29. John Freeman, *The Tyranny of E-Mail: The Four-Thousand-Year Journey to Your Inbox* (New York: Scribner, 2009).

30. Larry Rosen and Michelle Weil, *TechnoStress: Coping with Technology @Work @Home @ Play* (New York: John Wiley & Sons, 1997).

31. Mark Bauerlein, *The Dumbest Generation: How the Digital Age Stupefies Young Americans and Jeopardizes Our Future* (New York: Penguin, 2008).

32. David Trend, *The End of Reading: From Gutenberg to Grand Theft Auto* (New York: Peter Lang, 2010).

33. Mark Helprin, *Digital Barbarism: A Writer's Manifesto* (New York: Collins, 2009).

34. Andrew Keen, *Digital Vertigo: How Today's Online Social Revolution Is Dividing, Diminishing, and Disorienting Us* (New York: St. Martin's Press, 2012).

35. Foer, *World without Mind*.

36. Frank Furedi, "Precautionary Culture and the Rise of Probabilistic Risk Assessment," *Erasmus Law Review* 2, no. 2 (2009).

37. Braden R. Allenby, *The Rightful Place of Science: Future Conflict & Emerging Technologies* (Tempe, AZ: Consortium for Science, Policy & Outcomes, 2016), p. 48.

38. Jane K. Cramer and A. Trevor Thrall, "Introduction: Understanding Threat Inflation," in *American Foreign Policy and the Politics of Fear: Threat Inflation Since 9/11*, ed. A. Trevor Thrall and Jane K. Cramer (Routledge, 2009), p. 1.

39. Steven Cerier, "Frankenfoods? A 'Terrible Word' That Could Describe More Foods Than You Might Realize," Genetic Literacy Project, April 13, 2018; See also Juma, *Innovation and Its Enemies*, pp. 224–56.

40. Philip M. Fernbach et al., "Extreme Opponents of Genetically Modified Foods Know the Least but Think They Know the Most," *Nature Human Behaviour* 3 (2019): 251–56.

41. Ian Sample, "Strongest Opponents of GM Foods Know the Least but Think They Know the Most," *The Guardian*, January 14, 2019.

42. Ilya Somin, "Survey Shows Most Extreme Opponents of GMO Foods Know the Least—Yet Think They Know the Most," Volokh Conspiracy, Reason website, January 15, 2019.

43. Jeffrey Smith, *Seeds of Deception: Exposing Industry and Government Lies about the Safety of Genetically Engineered Foods You're Eating* (Fairfield, IA: Yes Books, 2003).

44. Steven M. Druker, *Altered Genes, Twisted Truth: How the Venture to Genetically Engineer Our Food Has Subverted Science, Corrupted Government, and Systematically Deceived the Public* (Salt Lake City, UT: Clear River Press, 2015).

45. Emily Steel and Julia Angwin, "MySpace Receives More Pressure to Limit Children's Access to Site," *Wall Street Journal*, June 23, 2006.

46. Deleting Online Predators Act of 2006, H.R. 5319, 109th Cong. (2006). See also Adam Thierer, "The Middleman Isn't the Problem," Philly.com, May 31, 2006.

47. Adam Thierer, "Would Your Favorite Website Be Banned by DOPA?," *Technology Liberation Front* (blog), March 10, 2007.

48. Adam Thierer, "A Framework for Responding to Online Safety Risks," in *Minding Minors Wandering the Web: Regulating Online Child Safety*, ed. Simone van der Hof, Bibi van den Berg, and Bart Schermer (The Hague, Neth.: Springer, 2014), pp. 39–66.

49. Adam Thierer, "Technopanics, Threat Inflation, and the Danger of an Information Technology Precautionary Principle," *Minnesota Journal of Law, Science & Technology* 14, no. 1 (2013): 312–50.

50. Richard A. Serrano, "Cyber Attacks Seen as a Growing Threat," *Los Angeles Times*, February 11, 2011, p. A18 ("[T]he potential for the next Pearl Harbor could very well be a cyber attack.").

51. Harry Raduege, "Deterring Attackers in Cyberspace," *The Hill*, September 23, 2011.

52. David Kravets, "Vowing to Prevent 'Cyber Katrina,' Senators Propose Cyber Czar," *Wired*, April 1, 2009.

53. Kate Fazzini, "Power Outages, Bank Runs, Changed Financial Data: Here Are the 'Cyber 9/11' Scenarios That Really Worry the Experts," *CNBC*, November 18, 2018, https://www.cnbc.com/2018/11/18/cyber-911-scenarios-power-outages-bank-runs-changed-data.html.

54. Joe Weiss, "Industrial Control Systems: The Holy Grail of Cyberwar," *Medium*, March 24, 2017.

55. Richard A. Clarke and Robert K. Knake, *Cyber War: The Next Threat to National Security and What to Do About It* (New York: ECCO, 2010).

56. Bruce Schneier, *Click Here to Kill Everybody: Security and Survival in a Hyper-connected World* (New York: W. W. Norton, 2018).

57. Giovanni Buttarelli, "Closing Remarks: CPDP 2019," Conference on Computers, Privacy, and Data Protection, February 1, 2019.

58. James Heim, *Voluntary Enslavement: Technology's Fast Development Reduces Diversity and Freedom* (self-pub., 2016).

59. Richard A. Posner, *Catastrophe: Risk and Response* (Oxford: Oxford University Press, 2004), p. 100.

60. Jill Lepore, "A Golden Age for Dystopian Fiction," *New Yorker*, June 5 and 12, 2017.

61. Jeff Jarvis, "Optimism Doesn't Sell," *Medium,* May 25, 2015.

62. Matt Ridley, *The Rational Optimist: How Prosperity Evolves* (New York: Harper Collins, 2010), p. 294.

63. Dan Wang, "Definite Optimism as Human Capital," *Dan Wang* (blog), August 7, 2017.

64. Steven Pinker, *Enlightenment Now: The Case for Reason, Science, Humanism, and Progress* (New York: Viking, 2018), p. 49. See also Mitch Daniels, "Things Really Aren't That Bad. But We Like to Think They Are," *Washington Post*, December 26, 2018.

65. Edward W. Lawless, *Technology and Social Shock* (New Brunswick, NJ: Rutgers University Press, 1977), p. 530.

66. Matt Ridley, "Why Is It So Cool to Be Gloomy?," *Wall Street Journal*, November 16, 2018.

67. Clive Thompson, *Smarter Than You Think: How Technology Is Changing Our Minds for the Better* (New York: Penguin Press, 2013), p. 283. See also Ridley, *The Rational Optimist*, p. 280.

68. Quoted in Ridley, "Why Is It So Cool to Be Gloomy?"

69. Robin Hanson, *The Age of Em* (Oxford: Oxford University Press, 2016), p. 41.

70. Brady Gerber, "Dystopia for Sale: How a Commercialized Genre Lost Its Teeth," Literary Hub, February 8, 2018.

71. Adam Thierer, "Is It 'Techno-Chauvinist' & 'Anti-Humanist' to Believe in the Transformative Potential of Technology?," *Medium*, September 18, 2018; Florman, *The Existential Pleasures*, p. 72.

72. Juma, *Innovation and Its Enemies*, p. 309.

73. Juma, p. 309.

74. Hanson, *The Age of Em*, p. 41.

75. Noel Kingsbury, *Hybrid: The History and Science of Plant Breeding* (Chicago: University of Chicago Press, 2009), p. 6.

76. Val Giddings, "Gene Editing, GMOs, and Fear-Based Marketing," Innovation Technology and Innovation Foundation (ITIF), Innovation Files, June 18, 2018; Mark Lynas, "Confession of an Anti-GMO Activist," *Wall Street Journal*, June 22, 2018.

77. Nicola Davis, "Weird New Fruits Could Hit Aisles Soon Thanks to Gene-Editing," *The Guardian*, July 19, 2018; Jacob Bunge, "Sizzling Steaks May Soon Be Lab-Grown," *Wall Street Journal*, February 1, 2016; Jacob Bunge, "Startup Producing Cell-Grown Meat Raises New Funding," *Wall Street Journal*, July 16, 2018.

78. Andrew Maynard, Sheril Kirshenbaum, and Walter Johnson, "Would You Eat 'Meat' from a Lab? Consumers Aren't Necessarily Sold on 'Cultured Meat,'" *Singularity Hub*, August 31, 2018.

79. Callan Boys, "The Future of Food: Lab-Grown Meat and 3D-Printed Meals," *Stuff*, June 23, 2015, https://www.stuff.co.nz/life-style/food-wine/69606846/The-future-of-food-Lab-grown-meat-and-3D-printed-meals; Joi Ito, "Fake Meat, Served Six Ways," *Wired*, July 2, 2018.

80. Robert D. Atkinson, "A Brief History of Our Allegedly Doomed Technological Future," *Medium*, July 13, 2018.

81. David H. Autor, "Why Are There Still So Many Jobs? The History and Future of Workplace Automation," *Journal of Economic Perspectives* 29, no. 3 (Summer 2015): 3.

82. Adam Thierer, *Permissionless Innovation: The Continuing Case for Comprehensive Technological Freedom*, 2nd ed. (Arlington, VA: Mercatus Center at George Mason University, 2016), p. 100.

83. Bret Swanson, "Automation Creates Good Jobs," *U.S. Chamber of Commerce Foundation* (blog), November 5, 2018.

84. Adam Thierer, "Technopanics, Threat Inflation, and the Danger of an Information Technology Precautionary Principle," *Minnesota Journal of Law, Science & Technology* 14, no. 1 (2013), pp. 312–50; Adam Thierer, "Why Do We Always Sell the Next Generation Short?," *Forbes*, January 8, 2012.

85. Adam Thierer, "Are You an Internet Optimist or Pessimist? The Great Debate over Technology's Impact on Society," *Technology Liberation Front* (blog), January 31, 2010.

86. Elizabeth L. Eisenstein, *The Printing Press as an Agent of Change* (New York: Cambridge University Press, 1979).

87. Robert D. Atkinson, "'It's Going to Kill Us!' and Other Myths about the Future of Artificial Intelligence," Information Technology and Innovation Foundation, June 2016.

88. Hunter Oatman-Stanford, "Don't Panic: Why Technophobes Have Been Getting It Wrong since Gutenberg," *Collectors Weekly*, September 28, 2015.

89. Adam Thierer, "Muddling Through: How We Learn to Cope with Technological Change," *Medium*, June 30, 2014.

90. Richard C. Longworth, "A Misguided View on Technology and Humanism," Chicago Council on Global Affairs, *The Midwesterner* (blog), January 20, 2015.

91. Longworth, "Misguided View."

92. Otto Bettmann, *The Good Old Days—They Were Terrible!* (New York: Random House, 1974); Alan Jay Levinovitz, "It Was Never Golden," Aeon, August 17, 2016.

93. Pinker, *Enlightenment Now*, p. 79; Greg Ip, "The World Is Getting Quietly, Relentlessly Better," *Wall Street Journal*, January 2, 2019; Florman, *The Existential Pleasures*, p. 62.

94. Pinker, *Enlightenment Now*, p. 52.

95. Kingsbury, *Hybrid*, p. 6.

96. David F. Noble, *The Religion of Technology* (New York: Penguin Books, 1999); L. M. Sacasas, "Revisiting 'The Religion of Technology'," April 28, 2012, and "The Virgin and the Data Center," January 17, 2019, *Technology, Culture, and Ethics* (blog).

97. Ethan Wham, "Experimentation and Failure in Disruptive Innovation," DisCo Disruptive Competition Project, December 5, 2018.

98. Joel Mokyr, Chris Vickers, and Nicolas L. Ziebarth, "The History of Technological Anxiety and the Future of Economic Growth: Is This Time Different?," *Journal of Economic Perspectives* 29, no. 3 (Summer 2015): 31. See also Brink Lindsey, "Why Growth Is Getting Harder," Cato Institute Policy Analysis no. 737, October 8, 2013, p. 11, http://www.cato.org/publications/policy-analysis/why-growth-getting-harder.

99. Paul Dragos Aligica, *Prophecies of Doom and Scenarios of Progress: Herman Kahn, Julian Simon, and the Prospective Imagination* (New York: Continuum International Publishing Group, 2007), p. 74.

100. Donald J. Boudreaux, "What's Your Moonshot?," Mercatus Center at George Mason University, Mercatus Original Video, November 16, 2017. See also Tyler Cowen, *Stubborn Attachments: A Vision for a Society of Free, Prosperous, and Responsible Individuals* (San Francisco: Stripe Press, 2018).

101. W. Brian Arthur, *The Nature of Technology: What It Is and How It Evolves* (New York: Free Press, 2009), p. 10.

102. Cowen, *Stubborn Attachments*.

103. Arti Rai, Stuart Graham, and Mark Doms, "Patent Reform: Unleashing Innovation, Promoting Economic Growth, and Producing High-Paying Jobs," white paper, U.S. Department of Commerce, April 13, 2010, pp. 1–3. Echoing these findings, two major economic surveys similarly found that technological progress accounts for 30 to 34 percent of growth in Western countries; Scott L. Baier, Gerald P. Dwyer Jr., and Robert Tamura, "How Important Are Capital and Total Factor Productivity for Economic Growth?," *Economic Inquiry* 44, no. 1 (2006): 23–49.

104. James Broughel and Adam Thierer, "Technological Innovation and Economic Growth: A Brief Report on the Evidence," Mercatus Center at George Mason University, research paper, February 2019.

105. Hans Rosling, *Factfulness: Ten Reasons We're Wrong about the World—and Why Things Are Better Than You Think* (New York: Flatiron Books, 2018).

106. Pinker, *Enlightenment Now*; Steven Pinker, "Enlightenment Wars: Some Reflections on 'Enlightenment Now,' One Year Later," *Quillette*, January 14, 2019.

107. Robert Bryce, *Smaller Faster Lighter Denser Cheaper: How Innovation Keeps Proving the Catastrophists Wrong* (New York: Public Affairs, 2014).

108. Gregg Easterbrook, *It's Better Than It Looks: Reasons for Optimism in an Age of Fear* (New York: Public Affairs, 2018); Michael A. Cohen and Micah Zenko, *Clear and Present Safety: The World Has Never Been Better and Why That Matters to Americans* (New Haven, CT: Yale University Press, 2019); Bobby Duffy, *The Perils of Perception: Why We're Wrong about Nearly Everything* (London: Atlantic Books, 2018).

109. Bryce, *Smaller Faster Lighter Denser Cheaper*, pp. xxi–xxii.

110. Ip, "The World."

111. Rosling, *Factfulness*, p. 55.

112. Rosling, *Factfulness*, p. 52.

113. Rosling, *Factfulness*, pp. 52–53.

114. Max Roser, "Memorizing These Three Statistics Will Help You Understand the World," *Gates Notes* (blog), June 26, 2018.

115. Roser, "Memorizing."

116. Rosling, *Factfulness*, p. 63.

117. Kevin Kelly, *The Inevitable: Understanding the 12 Technological Forces That Will Shape Our Future* (New York: Viking, 2016), p. 165.

118. Rosling, *Factfulness*, p. 62.

119. Rosling, *Factfulness*, p. 62.

120. Rosling, *Factfulness*, p. 63.

121. Scott Wallsten, "Cloud Computing: Co-Invention for the Masses," *Technology Policy Institute* (blog), July 19, 2018.

122. Tim Harford, *50 Things That Made the Modern Economy* (New York: Riverhead Books, 2017).

123. Tim Harford, "Why the Falling Cost of Light Matters," BBC News, February 6, 2017.

124. Harford, "Falling Cost of Light."

125. Marian L. Tupy, "Market Has Achieved What Marx Wanted—Less Labor," HumanProgress.org, July 13, 2018.

126. Marian L. Tupy, "U.S. Cost of Living and Wage Stagnation, 1979–2015," HumanProgress.org, August 12, 2017.

127. Tim Bajarin, "The Smartphone Is the Swiss Army Knife of Gadgets," *Time*, November 18, 2013.

128. Juliette Garside, "Like an Elite Swiss Army Knife, the iPhone Is Now a Multi-Category Killer," *The Guardian*, January 10, 2012.

129. Marian L. Tupy, "Computers Allow Us to Accomplish More with Less," HumanProgress.org, September 20, 2016.

130. Aaron Wildavsky, *Searching for Safety* (New Brunswick, CT: Transaction Books, 1988), p. 103.

131. Ian Barbour, *Ethics in an Age of Technology* (New York: Harper One, 1993), p. 3.

132. Robert D. Friedel, *A Culture of Improvement: Technology and the Western Millennium* (Cambridge, MA: MIT Press, 2007), p. 1.

133. Joseph A. Schumpeter, "The Analysis of Economic Change," *Review of Economics and Statistics* 17, no. 4 (1935): 4.

134. Peter Thiel, *Zero to One: Notes on How to Build the Future* (New York: Crown Business, 2014).

135. Israel M. Kirzner and Frederic Sautet, "The Nature and Role of Entrepreneurship in Markets: Implications for Policy," *Mercatus Policy Series Policy Primer*, No. 4, Arlington, VA: Mercatus Center at George Mason University, June 2006; Arthur M. Diamond Jr., "Schumpeter's Creative Destruction: A Review of the Evidence," *Journal of Private Enterprise* 22, no. 1 (Fall 2006): 120–46.

136. H. W. Lewis, *Technological Risk* (New York: W. W. Norton, 1990), p. x.

137. Jerry Ellig and Daniel Lin, "A Taxonomy of Dynamic Competition Theories," in *Dynamic Competition and Public Policy: Technology, Innovation, and Antitrust Issues*, ed. Jerry Ellig (Cambridge: Cambridge University Press, 2001), p. 18.

138. William Baumol, *Entrepreneurship, Management, and the Structure of Payoffs* (Cambridge, MA: MIT Press, 1993).

139. Sander Wennekers and Roy Thurik, "Linking Entrepreneurship and Economic Growth," *Small Business Economics* 13 (February 1999): 27–55; Randall G. Holcombe, *Entrepreneurship and Economic Progress* (New York: Routledge, 2007); David B. Audretsch, Max C. Keilbach, and Erik E. Lehmann, *Entrepreneurship and Economic Growth* (Oxford: Oxford University Press, 2006); Benjamin Powell and Rick Weber, "Economic Freedom and Entrepreneurship: A Panel Study of the United States," *American Journal of Entrepreneurship* 6, no. 1 (June 2013): 64–84; Joshua Hall and Russell Sobel, "Freedom, Entrepreneurship, and Growth," in *Lessons from the Poor: Triumph of the Entrepreneurial Spirit*, ed. Alvaro Llosa (Oakland, CA: Independent Institute, 2008), pp. 247–68.

140. William Baumol, "Entrepreneurship in Economic Theory," *American Economic Review* 58, no. 2 (May 1968): 66.

141. Joe Carlen, *A Brief History of Entrepreneurship: The Pioneers, Profiteers, and Racketeers Who Shaped Our World* (New York: Columbia University Press, 2016), pp. 214–15.

142. Peter J. Boettke and Christopher J. Coyne, "Context Matters: Institutions and Entrepreneurship," *Foundations and Trends in Entrepreneurship* 5, no. 3 (2009): 159, 160.

143. Edd S. Noell, Stephen L. S. Smith, and Bruce G. Webb, *Economic Growth: Unleashing the Potential of Human Flourishing* (Washington: Values and Capitalism, 2013), pp. 2–3.

144. Benjamin M. Friedman, "The Moral Consequences of Economic Growth," *Society* 43, no. 2 (January/February 2006): 15.

145. Peter J. Boettke, *Living Economics: Yesterday, Today, and Tomorrow* (Oakland, CA: Independent Institute, 2012), p. 30.

146. Tyler Cowen, "The Case for the Longer Term," Cato Unbound, January 9, 2019, https://www.cato-unbound.org/2019/01/09/tyler-cowen/case -longer-term.

147. Rosling, *Factfulness,* p. 64.

148. Rosling, *Factfulness,* p. 64.

149. Arthur M. Diamond Jr., *Openness to Creative Destruction: Sustaining Innovative Dynamism* (Oxford: Oxford University Press, 2019), pp. 43–45; Nick Gillespie, "Tyler Cowen's Gospel of Prosperity," *Reason*, March 2019.

150. Emmanuel G. Mesthene, *Technological Change: Its Impact on Man and Society* (Cambridge, MA: Harvard University Press, 1970), p. 62.

151. Friedel, *A Culture of Improvement*, p. 543.

152. Brian Solis, "Innovation Is a Gift Worth Getting: Competing for the Future Starts with Letting Go of the Past," @ Brian Solis, October 19, 2016.

153. Arthur, *The Nature of Technology*, p. 16.

154. Subra Suresh, "Research Universities, Innovation, and Growth," *Research-Technology Management* 58, no. 6 (2015): 22.

155. Ridley, *The Rational Optimist*, pp. 257–58; Harford, *50 Things*, p. 166.

156. John Nosta, "The New Hierarchy of Needs: Food, Water, Shelter . . . Tech!" *Forbes*, October 31, 2017.

157. Mark Strauss, "Four-in-Ten Americans Credit Technology with Improving Life Most in the Past 50 Years," Pew Research Center, October 12, 2017.

158. Adam Thierer, "Deep Technologies & Moonshots: Should We Dare to Dream?," *Medium*, September 7, 2018.

159. For critiques of moonshots and "moonshot fallacies," see Walter D. Valdivia, "Are Moonshots Giant Leaps of Faith?," *Issues in Science and Technology* XXXIII, no. 3 (Spring 2017); Mark P. Mills, "Making Technological Miracles," *The New Atlantis: A Journal of Technology & Society* (Spring 2017): 37–55. Also, regarding the overuse of the term "disruption," see Joshua Gans, *The Disruption Dilemma* (Cambridge, MA: MIT Press, 2016), pp. vii, viii.

160. Boudreaux, "What's Your Moonshot?"

161. Swati Chaturvedi, "So What Exactly Is 'Deep Technology'?," LinkedIn, July 28, 2015.

162. Chaturvedi, "So What Exactly Is 'Deep Technology'?"

163. Christopher Freeman and Luc Soete, *Economics of Industrial Innovation*, 3rd ed. (London: Routledge, 1997), p. 312.

164. Ford, *Innovation and the State*, p. 167.

165. Adam Thierer, "Making the World Safe for More Moonshots," Mercatus Center at George Mason University, *The Bridge*, February 5, 2018.

166. Adam Thierer and Trace Mitchell, "The Many Forms of Entrepreneurialism," Mercatus Center at George Mason University, *The Bridge*, August 30, 2018.

167. Eric von Hippel, *Free Innovation* (Cambridge, MA: MIT Press, 2017), p. 1.

168. Eric von Hippel, *Sources of Innovation*, (Oxford: Oxford University Press, 1988).

169. Sonali K. Shah, Sheryl Winston Smith, and E. J. Reedy, "Who Are User Entrepreneurs? Findings on Innovation, Founder Characteristics, and Firm Characteristics," *The Kauffman Firm Survey*, Ewing Marion Kauffman Foundation, February 2012.

170. Virgil Storr, Stefanie Haeffele, and Laura Grube, "The Entrepreneur as a Driver of Social Change," in *Community Revival in the Wake of Disaster* (New York: Palgrave Macmillan, 2018), pp. 11–32.

171. James Austin, Howard Stevenson, and Jane Wei-Skillern, "Social and Commercial Entrepreneurship: Same, Different, or Both?," *Entrepreneurship Theory and Practice* 30, no. 1 (January 2006): 2.

172. Austin, "Social and Commercial Entrepreneurship," p. 3.

173. von Hippel, *Free Innovation*, p. 17. ("Free innovators sometimes have important practical, legal, and regulatory advantages over producers.")

174. Thierer, *Permissionless Innovation*, p. 44.

175. Thierer, *Permissionless Innovation*, p. 44.

176. Daniel Castro and Michael McLaughlin, "Ten Ways the Precautionary Principle Undermines Progress in Artificial Intelligence," Information Technology and Innovation Foundation, February 4, 2019.

177. See also von Hippel, *Free Innovation*, p. 128.

178. See, for example, Julia Powles and Helen Nissenbaum, "The Seductive Diversion of 'Solving' Bias in Artificial Intelligence," *Medium*, December 7, 2018.

179. Norman Lewis, "The Enduring Wisdom of the Crowd," *Spiked*, October 24, 2018.

180. Thierer, *Permissionless Innovation*, p. 2.

181. Peter J. Boettke, *Great Thinkers in Economics: F. A. Hayek* (London: Palgrave Macmillan, 2018), p. 112.

182. Martin Rees, *On the Future: Prospects for Humanity* (Princeton, NJ: Princeton University Press, 2018), p. 136.

183. Giandomenico Majone, "What Price Safety? The Precautionary Principle and Its Policy Implications," in *Risk Regulation in the European Union: Between Enlargement and Internationalization*, ed. Giandomenico Majone (Florence, Italy: European University Institute, Robert Schuman Centre for Advanced Studies, 2003), p. 49.

184. Cass R. Sunstein, *Laws of Fear: Beyond the Precautionary Principle* (Cambridge: Cambridge University Press, 2005), p. 16.

185. Majone, "What Price Safety?," pp. 49–50.

186. Ford, *Innovation and the State*, p. 191.

187. Jason Potts, "Innovation Policy in a Global Economy," *Journal of Entrepreneurship and Public Policy* 5, no. 3 (2016).

188. Michael Porter, *The Competitive Advantage of Nations* (New York: Free Press, 1990), p. 125.

189. Benjamin Netanyahu, "Innovation Nation," *The Economist: The World in 2018* (London: *The Economist*, 2017).

190. Sander Wennekers and Roy Thurik, "Linking Entrepreneurship and Economic Growth," *Small Business Economics* 13 (February 1999): 40.

191. Dustin Chambers and Jonathan Munemo, "The Impact of Regulations and Institutional Quality on Entrepreneurship," Mercatus Center at George Mason University, *Mercatus Research* (2017): 25. See also Steven Globerman and George Georgopoulos, "Regulation and the International Competitiveness of the U.S. Economy," Mercatus Center at George Mason University, working paper, September 18, 2012.

Chapter 2

1. Niklas Elert and Magnus Henrekson, "Evasive Entrepreneurialism," *Small Business Economics* 47 (2016): 96. See also Christopher J. Coyne and Peter T. Leeson, "The Plight of Underdeveloped Countries," *Cato Journal* 24, no. 3 (Fall 2004): 237.

2. Elizabeth Pollman and Jordan M. Barry, "Regulatory Entrepreneurship," *Southern California Law Review* 90 (2017): 392, 399.

3. Peter J. Boettke and Christopher J. Coyne, "Context Matters: Institutions and Entrepreneurship," *Foundations and Trends in Entrepreneurship* 5, no. 3 (2009): 180. See also Robert Dahl, *Who Governs?* (New Haven, CT: Yale University Press, 1961).

4. Thomas DiLorenzo, "Competition and Political Entrepreneurship," *Review of Austrian Economics* 2, no. 1 (1998): 59–71. See also Fred S. McChesney, *Money for Nothing: Politicians, Rent Extraction, and Political Extortion* (Cambridge, MA: Harvard University Press, 1997).

5. Randy T. Simmons, Ryan M. Yonk, and Diana W. Thomas, "Bootleggers, Baptists, and Political Entrepreneurs: Key Players in the Rational Game and Morality Play of Regulatory Politics," *The Independent Review* 15, no. 3 (Winter 2011): 370.

6. William Baumol drew a distinction between "productive" versus "unproductive" entrepreneurship in his work. He argued that "while the total supply of entrepreneurs varies among societies, the productive contribution of the society's entrepreneurial activities varies much more because of their allocation between productive activities such as innovation and largely unproductive activities such as rent seeking or organized crime. This allocation is heavily influenced by the relative payoffs society offers to such activities." William J. Baumol, "Entrepreneurship: Productive, Unproductive and Destructive," *Journal of Political Economy* 98, no. 5 (1990): 893.

7. Boettke and Coyne, "Context Matters": 163.

8. Valerie Bauerlein, "Pothole Vigilantes Fill the Streets, Plugging Gaps Left by City Workers," *Wall Street Journal*, May 30, 2018.

9. Joe Carlen, *A Brief History of Entrepreneurship: The Pioneers, Profiteers, and Racketeers Who Shaped Our World* (New York: Columbia University Press, 2016), p. 2.

10. George Gilder, *The Spirit of Enterprise* (New York: Simon and Schuster, 1984), p. 246.

11. Joseph Schumpeter, *Capitalism, Socialism and Democracy*, 3rd ed. (New York: Harper Perennial, 1942, 2008).

12. Israel Kirzner, *Competition and Entrepreneurship* (Chicago: University of Chicago Press, 1978); Israel Kirzner, "The Alert and Creative Entrepreneur: A Clarification," *Small Business Economics* 32, no. 2 (February 2009): 145–52.

13. Robert P. Murphy, "The Connection between Entrepreneurship and Economic Prosperity: Theory and Evidence," in *Demographics and Entrepreneurship: Mitigating the Effects of an Aging Population*, ed. Steven Globerman and Jason Clemons (Canada: Fraser Institute, 2018).

14. Calestous Juma, *Innovation and Its Enemies: Why People Resist New Technologies* (New York: Oxford University Press, 2016).

15. Juma, *Innovation and Its Enemies*, p. 315.

16. Diane Hamblen, "Only the Limits of Our Imagination: An Exclusive Interview with RADM Grace M. Hopper," *Ships Ahoy*, July 1986.

17. Hamblen, "Limits of Our Imagination," at 398.

18. Adam Thierer, *Permissionless Innovation: The Continuing Case for Comprehensive Technological Freedom*, 2nd ed. (Arlington, VA: Mercatus Center at George Mason University, 2016).

19. Pollman and Barry, "Regulatory Entrepreneurship": 7.

20. Pollman and Barry, "Regulatory Entrepreneurship": 4.

21. Pollman and Barry, "Regulatory Entrepreneurship": 7.

22. Rob Tracinski, "Civil Disobedience as a Business Model," *RealClearFuture*, April 12, 2017.

23. Stephanie Mehta, "Meet Uber's Political Genius," *Vanity Fair*, June 17, 2016.

24. Victor Fleischer, "Regulatory Arbitrage," *Texas Law Review* 89 (2010): 227–89.

25. Fleischer, "Regulatory Arbitrage."

26. Issie Lapowsky, "Uber's New Fake Feature in NYC Derides Regulators," *Wired*, July 16, 2015.

27. Kevin Roose, "Uber Just Added a 'De Blasio' Tab to Its App in New York to Fight the Mayor's Proposed Legislation," Splinter, July 16, 2015.

28. Fitz Tepper, "Uber Launches 'De Blasio's Uber' Feature in NYC with 25-Minute Wait Times," *TechCrunch*, July 16, 2015.

29. Jillian Jorgensen and Will Bredderman, "Bill de Blasio's Quest to Cap Uber Ends with a Whimper," *Observer*, January 15, 2016.

30. Jorgensen and Bredderman, "de Blasio's Quest to Cap Uber."

31. Tracinski, "Civil Disobedience."

32. Brad Stone, "The $99 Billion Idea," *Bloomberg Businessweek*, January 26, 2017.

33. Adam Thierer, "Regulatory Capture: What the Experts Have Found," *Technology Liberation Front* (blog), December 19, 2010.

34. Michael Farren, Christopher Koopman, and Matthew Mitchell, "Rethinking Taxi Regulations: The Case for Fundamental Reform," Mercatus Center at George Mason University, *Mercatus Research*, July 2016.

35. Farren, Koopman, and Mitchell, " Rethinking Tax Regulations," pp. 8–11.

36. Mark W. Frankena and Paul A. Pautler, "An Economic Analysis of Taxicab Regulation," *Bureau of Economics Staff Report*, Federal Trade Commission, Washington (May 1984), p. 1.

37. Christopher Mims, "Tech's Innovators Are Starting to Ask Permission, Rather Than Forgiveness," *Wall Street Journal*, July 19, 2018.

38. Katie Benner, "Airbnb Settles Lawsuit with Its Hometown, San Francisco," *New York Times*, May 1, 2017.

39. Matthew Yglesias, "Uber's Toxic Culture of Rule Breaking, Explained," *Vox,* March 21, 2017.

40. *Business Travel Blog*, "Uber and Lyft Have Their Bluff Called in Austin," *The Economist*, blog entry by Gulliver [pseud.], May 17, 2016.

41. Brian Jencunas, "Protecting Workers in New York City Means Protecting Ride-Sharing," *R Street* (blog), May 25, 2018.

42. James Doubek, "New York City Temporarily Halts More Uber and Lyft Cars on the Road," NPR, August 9, 2018.

43. Costas Pitas, "'Unfit' Uber Stripped of London License, CEO Tweets 'pls work w/us'," *Reuters*, September 22, 2017.

44. Arjun Kharpal and Sara Salinas, "Uber Allowed to Operate in London Again After Judge Overturns Ban," CNBC, June 26, 2018.

45. Ted Graham, *The Uber of Everything: How the Freed Market Economy Is Disrupting and Delighting* (Middletown, DE: All Above Press, 2017).

46. Evan Burfield with J. D. Harrison, *Regulatory Hacking: A Playbook for Startups* (New York: Portfolio/Penguin, 2018); Bradley Tusk, *The Fixer: My Adventures Saving Startups from Death by Politics* (New York: Portfolio/Penguin, 2018).

47. Jonathan A. Knee, "Review: Why Start-Ups Need a Regulatory Strategy to Succeed," *New York Times*, September 11, 2018.

48. Robert F. Graboyes and Jordan Reimschisel, "Opening the Door for Medical Innovation," *U.S. News*, April 24, 2017.

49. "Warning Letters: 23andMe, Inc.," Alberto Gutierrez to Anne Wojcicki, FDA.gov, November 22, 2013.

50. Adam Thierer, "Innovation Arbitrage, Technological Civil Disobedience & Spontaneous Deregulation," *Medium*, December 7, 2016.

51. Elert and Henrekson, "Evasive Entrepreneurialism": 102.

52. Mario Cervantes and Dominique Guellec, "The Brain Drain: Old Myths, New Realities," *OECD Observer*, no. 230 (January 2002).

53. Anupam Chander, *The Electronic Silk Road: How the Web Binds the World Together in Commerce* (New Haven, CT: Yale University Press, 2013), p. 5. See also Stephen J. Kobrin, "Back to the Future: Neomedievalism and the Postmodern Digital World Economy," in *Globalization and Governance*, ed. Aseem Prakash and Jeffrey A. Hart (London and New York: Routledge, 1999), pp. 165–87.

54. Milton Friedman, "The New Economic Order," Hong Kong Centre for Economic Research, *HKCER Letters* 23 (November 1993).

55. Jason Potts, "Innovation Policy in a Global Economy," *Journal of Entrepreneurship and Public Policy* 5, no. 3 (2016): 309.

56. Richard Baldwin, *The Great Convergence: Information Technology and the New Globalization* (Cambridge, MA: Belknap Press of Harvard University Press, 2016), p. 175.

57. Madsen Pirie, "Game-Changers," *Adam Smith Institute* (blog), July 24, 2018.

58. James Dale Davidson and William Rees-Mogg, *The Sovereign Individual: Mastering the Transition to the Information Age* (New York: Simon & Schuster, 1999), p. 202.

59. Alfred C. Aman Jr., "Administrative Law for a New Century," in *The Province of Administrative Law*, ed. Michael Taggart (Oxford: Hart Publishing, 1997), p. 102.

60. Aman Jr., "Administrative Law for a New Century."

61. Steven Millward, "How Big Is China's Tech Industry? Here Are the Latest Stats," Techinasia, July 26, 2018.

62. Michael Newell, "Top 5 Tech Hubs to Rival Silicon Valley," *The New Economy*, January 18, 2019.

63. Ed Pilkington, "Amazon Tests Delivery Drones at Secret Canada Site After US Frustration," *The Guardian*, March 30, 2015; Ruth Reader, "Amazon Spurns Slow FAA, Reveals It's Been Testing Drones Abroad," *VentureBeat*, March 24, 2015.

64. Alan McQuinn, "Commercial Drone Companies Fly Away from FAA Regulations, Go Abroad," *Inside Sources*, September 30, 2014.

65. Christopher Koopman, "The Risk of Risk Aversion at the Federal Aviation Administration," *The Hill*, August 11, 2018.

66. National Academies of Sciences, Engineering, and Medicine, *Assessing the Risks of Integrating Unmanned Aircraft Systems (UAS) into the National Airspace System: Consensus Study Report* (Washington: The National Academies Press, 2018), p. S-2.

67. National Academies of Sciences, Engineering, and Medicine, *Assessing the Risks.* See also Adam Thierer, "National Academies Report Rips FAA's Risk-Averse Regulatory Culture," *Technology Liberation Front* (blog), June 12, 2018.

68. Andrea O'Sullivan and Michael Kotrous, "The Good and the Bad of FAA Reauthorization Drone Policy," Mercatus Center at George Mason University, *The Bridge*, November 6, 2018.

69. Jiayang Fan, "How E-Commerce Is Transforming Rural China," *New Yorker*, July 23, 2018.

70. Pilkington, "Amazon Tests Delivery Drones."

71. Kate Baggaley, "Drones Are Fighting Wildfires in Some Very Surprising Ways," Mach, November 16, 2017.

72. Malek Murison, "Search and Rescue Police Drone Finds Woman with Dementia," WeTalkUAV, November 9, 2017.

73. Ella Koscher, "Can Flying Drones Save Whales Trapped in Fishing Gear?," Mach, June 12, 2018.

74. Evan Ackerman, "Useful and Timely Delivery Drone Drops Life Preserver to Australian Swimmers," *IEEE Spectrum,* January 22, 2018.

75. Esther Landhuis, "Medical Cargo Could Be the Gateway for Routine Drone Deliveries," NPR, March 10, 2018.

76. "Public Safety Drones Save Four Lives in One Day," DJI, June 6, 2018; Jon Fingas, "Drone Fleets Could Find Lost Hikers in Forests without Using GPS," *Engadget*, November 4, 2018.

77. Marco Margaritoff, "Trump Administration Expands Drone Use to Beyond Visual Line of Sight," The Drive, October 25, 2017.

78. "UAS Integration Pilot Program Selection Announcement," remarks prepared for delivery by U.S. Secretary of Transportation Elaine L. Chao, May 9, 2018.

79. Larry Downes and Paul Nunes, "Regulating 23andMe to Death Won't Stop the New Age of Genetic Testing," *Wired*, January 1, 2014.

80. Samuel Gibbs, "DNA-Screening Test 23andMe Launches in UK After US Ban," *The Guardian*, December 1, 2014.

81. Jessica Firger, "U.K. Approves Sales of 23andMe Genetic Test Banned in U.S.," CBS News, December 3, 2014.

82. Andre Choulika, "The West Is Losing the Gene Editing Race. It Needs to Catch Up," *First Opinion* newsletter, *STAT*, October 29, 2018.

83. Kevin Potter and John Kepcha, "States Use Credits & Incentives to Attract Startups and Technology Companies," Credits & Incentives Talk with Deloitte, May 2015.

84. Adam Thierer, "Film Industry Tax Incentive Race to the Bottom Continues," *Technology Liberation Front* (blog), January 30, 2014; John A. Dove and Daniel Sutter, "Is There a Tradeoff between Economic Development Incentives and Economic Freedom? Evidence from the U.S. States," working paper, Mercatus Center at George Mason University, Arlington, VA, 2017; Peter T. Calcagno and Frank L. Hefner, "Targeted Economic Incentives: An Analysis of State Fiscal Policy and Regulatory Conditions," working paper, Mercatus Center at George Mason University, Arlington, VA, 2018.

85. Ryan Randazzo, "Arizona Getting Ahead of Autonomous Vehicle Industry by Stepping Aside," *AZ Central*, June 23, 26, 2017.

86. Arian Campo-Flores, "Cities Rush to Build Infrastructure—for Self-Driving Cars," *Wall Street Journal,* November 9, 2017.

87. Ann Thompson, "ODOT Wants to Make Ohio Even More Appealing to Self-Driving Car Industry," WOSU Public Media, March 13, 2017.

88. Jonathan Shieber, "California DMV Changes Rules to Allow Testing and Use of Fully Autonomous Vehicles," *TechCrunch*, October 11, 2017.

89. Julia Carrie Wong, "Uber Packs Up Failed Self-Driving Car Trial in California and Moves to Arizona," *The Guardian*, December 22, 2016.

90. Wong, "Uber Packs Up."

91. Nick Statt, "Uber Suspended from Autonomous Vehicle Testing in Arizona Following Fatal Crash," *The Verge*, March 26, 2018.

92. Andrew J. Hawkins, "Uber's Self-Driving Cars Return to Public Roads for the First Time Since Fatal Crash," *The Verge*, December 20, 2018; Sean Hollister, "Uber Won't Be Charged with Fatal Self-Driving Crash, Says Prosecutor," *The Verge*, March 5, 2019.

93. Carolyn Said and Benny Evangelista, "San Francisco to Robots: Don't Crowd Our Sidewalks," *San Francisco Chronicle*, December 6, 2017.

94. Melia Robinson, "Tech Founders Take Their Self-Driving Food-Delivery Robots Out of San Francisco to Focus on Cities Where They Feel More Welcome," *Business Insider*, May 9, 2018.

95. Eleonore Pauwels, "How China Is Pulling Ahead on AI and Biotech," *Axios*, December 3, 2017.

96. Jonah Lehrer, "Why Did Kobe Go to Germany?," *Grantland* (blog), April 16, 2012.

97. Lehrer, "Why Did Kobe Go to Germany?"

98. Rob Stein, "Clinic Claims Success in Making Babies with 3 Parents' DNA," NPR, June 6, 2018.

99. Marilynn Marchione and Christina Larson, "Could Anyone Have Stopped Gene-Edited Babies Experiment?," AP News, December 2, 2018.

100. Antonio Regalado, "Chinese Scientists Are Creating CRISPR Babies," *MIT Technology Review*, November 25, 2018.

101. Mark Harris, "FCC Accuses Stealthy Startup of Launching Rogue Satellites," *IEEE Spectrum*, March 9, 2018.

102. Jon Kelvey, "Pirate Radio in Space," *Slate*, April 9, 2018.

103. Anthony Serafini, Chief, Experimental Licensing Branch, FCC, to Sara Spangelo, Swarm Technologies, Inc., December 12, 2017. https://apps.fcc.gov/els/GetAtt.html?id=214769&x=

104. Experimental Special Temporary Authorization to Swarm Technologies, Inc., August 24, 2018–February 24, 2019, https://apps.fcc.gov/els/GetAtt.html?id=214769&x=.

105. Stephen Shankland, "Big Day for Drones as U.S. Endorses Tests of Package Delivery and More," CNet, May 9, 2018.

106. O'Sullivan and Kotrous, "The Good and the Bad."

107. Braden Allenby, "The Dynamics of Emerging Technology Systems," in *Innovative Governance Models for Emerging Technologies*, ed. Gary E. Marchant et al. (Northampton, MA: Edward Elgar, 2013), p. 33.

108. Department of Commerce, Bureau of Industry and Security, "Review of Controls for Certain Emerging Technologies," *Federal Register*, November 19, 2018, pp. 58201–2.

109. Adam Thierer and Jennifer Huddleston Skees, "Emerging Tech Export Controls Run Amok," *Technology Liberation Front* (blog), November 28, 2018.

110. Cade Metz, "Curbs on A.I. Exports? Silicon Valley Fears Losing Its Edge," *New York Times*, January 1, 2019.

111. Metz, "Curbs on A.I. Exports?"

112. Frost & Sullivan, "Wearable Technologies and Healthcare: Differentiating the Toys and Tools for 'Quantified-Self' with Actionable Health Use Cases," *Frost Perspectives* (blog), September 18, 2017.

113. Eric Topol, *The Creative Destruction of Medicine: How the Digital Revolution Will Create Better Care* (New York: Basic Books, 2012), p. 260.

114. Food and Drug Administration, Digital Health Innovation Action Plan, 2017. Note that the agency had "focused our oversight on mobile medical apps to only those that present higher risk to patients, while choosing not to enforce compliance for lower risk mobile apps."

115. Digital Health Innovation Action Plan, 2017.

116. Digital Health Innovation Action Plan, 2017, p. 1.

117. Food and Drug Administration, "Digital Health Software Precertification (Pre-Cert) Program," last updated April, 26, 2018.

118. Amanda Pedersen, "FDA Wants to Change the De Novo Pathway," MD+DI, December 4, 2018.

119. Asif Dhar, Mike Delone, and Dan Ressler, "Reimagining Digital Health Regulation: An Agile Model for Regulating Software in Health Care," Deloitte, 2018.

120. Robert Graboyes and Sara Rogers, "As Free Innovation Encounters Health Care Regulation, Think 'Soft Laws,'" *STAT*, September 12, 2018; Jordan Reimschisel, "Evolving Oversight of Digital Health," *Medium*, June 21, 2018.

121. Jordan Reimschisel, "Technology Could Enable Personal Medicine Whether We Like It or Not," *Medium*, August 10, 2017.

122. Tarun Wadhwa, "The Digitalization of Prosthetics Is Transforming How Wounded Service Members and Veterans Recover," *Singularity Hub*, Singularity Education Group, February 4, 2016.

123. Gina Martinez, "3D-Printed Guns Are Unchecked and Untraceable. And a Judge Blocked Them at the Last Minute," *Time*, July 31, 2018.

124. Adam D. Thierer and Adam Marcus, "Guns, Limbs, and Toys: What Future for 3D Printing?," *Minnesota Journal of Law, Science & Technology* 17, no. 2 (2016): 805–54.

125. Enabling the Future website, http://enablingthefuture.org/about.

126. "Father Helps Son with Cerebral Palsy Walk with Custom 3D Printed Orthosis," (blog entry on Formlabs corporate website), accessed June 27, 2018, https://formlabs.com/blog/father-helps-son-with-cerebral-palsy-walk-with-custom-3d-printed-orthosis.

127. Robert Graboyes, "How to Print Yourself a New Hand," CNN, October 24, 2014.

128. Wadhwa, "The Digitalization of Prosthetics."

129. J. D. Tuccille, "You'll Soon Be Able to Manufacture Anything You Want and Governments Will Be Powerless to Stop It," *Reason*, June 12, 2018.

130. Vanessa Romo, "Attorneys General Sue Trump Administration to Block 3D-Printed Guns," NPR, July 30, 2018.

131. Erica Goldberg, "3D Printable Guns as Free Speech?," *In a Crowded Theater* (blog), August 1, 2018; Andrea O'Sullivan, "When Code Is Speech, Tech Like 3D-Printed Guns Sees Greater Protection from Censorship," *Reason*, July 18, 2018.

132. Andrea O'Sullivan and Adam Thierer, "3D Printers, Evasive Entrepreneurs and the Future of Tech Regulation," Mercatus Center at George Mason University, *The Bridge*, August 1, 2018; Zusha Elinson, "Gun Advocates to Post 3D Weapon Plans on New Site," *Wall Street Journal*, August 1, 2018.

133. Christina Caron, "'Ghost Guns,' Homemade and Untraceable, Face Growing Scrutiny," *New York Times*, November 27, 2017.

134. Mark Tallman, "Our 3D-Printed Future," *Calibre Press*, September 5, 2017.

135. Dan Gillmor, *We the Media: Grassroots Journalism By the People, For the People* (Sebastopol, CA: O'Reilly Media, Inc.: July 2004).

136. Daniel Grushkin, "Biohackers Are About Open Access to Science, Not DIY Pandemics; Stop Misrepresenting Us," *STAT*, June 4, 2018.

137. Eleonore Pauwels and Sarah W. Denton, *The Rise of the New Bio-Citizen* (Washington: Wilson Center, 2018), p. 4.

138. For more information, see http://www.nightscout.info.

139. Kate Linebaugh, "Tech-Savvy Families Use Home-Built Diabetes Device," *Wall Street Journal*, May 9, 2016.

140. Naomi Kresge and Michelle Cortez, "The $250 Biohack That's Revolutionizing Life with Diabetes," *Bloomberg Businessweek*, August 8, 2018.

141. Sonya Collins, "#WeAreNotWaiting," *Genome* magazine, April 3, 2018.

142. Collins, "#WeAreNotWaiting."

143. Collins, "#WeAreNotWaiting."

144. Spencer Macnaughton and Conall Jones, "High Insulin Prices Drive Diabetics to Take Extreme Measures," WSJ Video, December 3, 2018.

145. Kresge and Cortez, "The $250 Biohack."

146. Matt McFarland, "A College Kid Spends $60 to Straighten His Own Teeth. What Could Possibly Go Wrong?," *Washington Post*, March 30, 2016.

147. Amos Dudley, "Orthoprint, or How I Open-Sourced My Face," *Squintin', lookin', doin'* (blog), March 10, 2016.

148. Tara C. Smith, "Why DIY Braces Are Actually a Terrible, Terrible Idea," *Washington Post*, March 25, 2016.

149. American Teledentistry Association, "Teleorthodontics and Access to Care: At Least 84 Percent of Underserved U.S. Counties Now Have Access to Orthodontic Treatment," news release, Markets Insider, June 26, 2018.

150. Kayla Stetzel, "Start-Ups Make Cheap Alternative to Braces: Dental Trade Groups Cry for Regulation," *Reason Hit & Run* (blog), May 2, 2018.

151. Jordan Reimschisel, "Biohackerspaces," *Medium*, June 29, 2017.

152. Reimschisel, "Biohackerspaces."

153. Bohyun Kim, "Biohackerspace, DIYbio, and Libraries," ACRL TechConnect, February 10, 2015.

154. Jenny McGruther, "Four Thieves Vinegar," Nourished Kitchen, July 19, 2011.

155. "Four Thieves Vinegar: Evolution of a Medieval Medicine," Plague Sage, accessed July 17, 2019, PlagueSage.com.

156. Daniel Oberhaus, "Meet the Anarchists Making Their Own Medicine," Motherboard, July 26, 2018, https://motherboard.vice.com /en_us/article/43pngb/how-to-make-your-own-medicine-four-thieves -vinegar-collective.

157. Four Thieves Vinegar, "Our Mission," accessed November 1, 2018, https://fourthievesvinegar.org/our-mission.

158. Oberhaus, "Meet the Anarchists."

159. Dan Stanton, "U.S. FDA Warns Against $30 DIY Alternative to Mylan's EpiPen," in-Pharma, September 21, 2016.

160. Kirsten V. Brown, "I Spent a Weekend with Cyborgs, and Now I Have an RFID Implant I Have No Idea What to Do With," Gizmodo, April 29, 2018.

161. Brown, "I Spent a Weekend with Cyborgs."

162. Zachary Kussin, "'Cyborg' Who Implanted Train Card in Hand Gets Fined for Not Having Ticket," *New York Post*, March 16, 2018.

163. Maddy Savage, "Thousands of Swedes Are Inserting Microchips Under Their Skin," NPR, October 22, 2018.

164. Emily Baumgaertner, "As D.I.Y. Gene Editing Gains Popularity, 'Someone Is Going to Get Hurt,'" *New York Times*, May 14, 2018.

165. Amy Dockser Marcus, "Scientists Call for Moratorium to Block Gene-Edited Babies," *Wall Street Journal*, March 13, 2019.

166. Emily Mullin, "Before He Died, This Biohacker Was Planning a CRISPR Trial in Mexico," *MIT Technology Review*, May 4, 2018.

167. Denise Grady and Sheila Kaplan, "F.D.A. Moves to Stop Rogue Clinics from Using Unapproved Stem Cell Therapies," *New York Times*, May 9, 2018.

168. Adam Thierer and Jordan Reimschisel, "Biohacking, Democratized Medicine, and Health Policy," Mercatus Center at George Mason University, *The Bridge*, July 19, 2019; Patricia J. Zettler, Christi J. Guerrini, and Jacob S. Sherkow, "Regulating Genetic Biohacking," *Science* 365, Issue 6448 (July 5, 2019): 34–6.

169. Sarah Finch, "FoodTech Revolution: Automation, Delivery and Convenience," Disruption, June 2018. See also Bret Swanson, "Innovation in the Economy's Oldest Sector—Food," *U.S. Chamber of Commerce Foundation* (blog), August 3, 2017.

170. Baylen Linnekin and Michael Bachmann, "The Attack on Food Freedom," Institute for Justice, May 2014.

171. Samantha Bomkamp, "Long-Simmering Food Truck Fight Heats Up on Chicago Streets, in Court," *Chicago Tribune*, October 21, 2016.

172. J. Justin Wilson and Robert Frommer, "Opportunity Lost," Institute for Justice, October 2016.

173. Bonnie Kristian, "These Food Truck Regulations Are Onerous, Unfair, and Outright Stupid," Rare, October 31, 2016.

174. Jennifer McDonald, *Flour Power: How Cottage Food Entrepreneurs Are Using Their Home Kitchens to Become Their Own Bosses*" (Arlington, VA: Institute for Justice, December 2017), p. 3.

175. Nina W. Tarr, "Food Entrepreneurs and Food Safety Regulation," *Journal of Food Law & Policy* 7 (2011): 36.

176. Adam Thierer and Trace Mitchell, "The Many Flavors of Food Entrepreneurialism," *The Bridge*, Mercatus Center at George Mason University, Arlington, VA, September 26, 2018.

177. Tarr, "Food Entrepreneurs": 56.

178. McDonald, *Flour Power*.

179. David Moye, "Throw a Potluck in Arizona and You Could Be Thrown in Jail," *Huffington Post*, February 18, 2016.

180. Krista Kafer, "Regulators Wore Us Down," *Denver Post*, July 3, 2014.

181. Cirrus Wood, "Private Chef Network 'Josephine' Closing," *Hoodline*, February 9, 2018.

182. Leilani Clark, "A New Law Would Legalize Selling Home-Cooked Food in California," *Civil Eats*, May 1, 2017.

183. Mary Bergin, "Home Bakers in Wisconsin Are Happily Selling Their Treats, Now That It's Legal—You Could Too," *Milwaukee Journal Sentinel*, February 26, 2019.

184. A. J. Jacobs, "Dinner Is Printed," *New York Times*, September 21, 2013; Jasper L. Tran, "3D-Printed Food," *Minnesota Journal of Law, Science & Technology* 17, no. 2 (2016).

185. Eli Dourado, "The Next Internet-Like Platform for Innovation? Airspace (Think Drones)," *Wired*, April 23, 2013.

186. Konstantin Kakaes, "Flying a Drone Can Be an Act of Civil Disobedience," *Washington Post*, April 15, 2015.

187. Jack Nicas, "Drone Ban? Corporations Skirt Rules," *Wall Street Journal*, February 19, 2015; Matt McFarland, "Drones: The Next Big Thing in Wedding Photography, or a Tacky Intrusion?," *Washington Post*, February 24, 2015.

188. Dan Friedman, "Rep. Sean Patrick Maloney Used Drone at His Wedding—Violating Federal Aviation Administration Rules," *Daily News*, July 14, 2014.

189. Federal Aviation Administration, "DOT and FAA Finalize Rules for Small Unmanned Aircraft Systems," press release, June 21, 2016.

190. Haye Kesteloo, "DJI Reports 65 People Rescued by Drones Last Year and 124 in Total Around the World," Drone DJ, April 30, 2018.

191. Larry Greenemeier, "Could Samaritan Drone Aircraft Help Hurricane Harvey Rescuers?," *Scientific American*, September 1, 2017.

192. Gregory S. McNeal, "FAA Admits That They Shouldn't Be Ordering People to Delete Drone Videos," *Forbes*, April 10, 2015.

193. Louise Roug, "Media Companies Challenge FAA Drone Ban," *Columbia Journalism Review*, May 7, 2014.

194 "U.S. Approved Ferguson No-Fly Area to Block Media," *USA Today*, November 2, 2014.

195. Peter Sachs, "Why Is There a TFR over Standing Rock?," *Drone Law Journal*, November 27, 2016.

196. Jason Koebler, "The Government Is Using a No Fly Zone to Suppress Journalism at Standing Rock," Motherboard, November 30, 2016, https://motherboard.vice.com/en_us/article/yp3kak/the-government-is-using-a-no-fly-zone-to-suppress-journalism-at-standing-rock.

197. Margot E. Kaminski, "Up in the Air: The Free-Speech Problems Raised by Regulating Drones," *Slate*, November 25, 2014.

198. John Goglia, "FAA Reverses Course, Grants Drone Journalist Permission to Fly in No-Fly Zone over Standing Rock," *Forbes*, December 2, 2016.

199. Eliza Strickland, "Drone Delivery Becomes a Reality in Remote Pacific Islands," *IEEE Spectrum*, May 31, 2018.

200. "Drone Delivers Abortion Pills to Northern Irish Women," *The Guardian,* June 21, 2016.

201. Anthony R. Fellow, *American Media History* (Belmont, CA: Wadsworth Group, 2005), p. 59.

202. Jennifer Huddleston and Trace Mitchell, "Transportation 3.0," *The Bridge*, Mercatus Center at George Mason University, Arlington, VA, August 29, 2018; Vinod Khosla, "Reinventing Transportation," *Medium*, January 6, 2018.

203. Christian Britschgi, "The War on Waze," *Reason*, April 25, 2018.

204. Fidel Martinez, "Cops Want Waze to Get Rid of Its Police-Tracking Feature," Splinter, January 26, 2015.

205. "The New Waze Carpool," Waze (blog), March 4, 2018.

206. Eliot Brown, "Adults Are Terrorizing San Francisco on Tiny Electric Scooters," *Wall Street Journal*, April 25, 2018.

207. Andrew J. Hawkins, "How Bird Plans to Spread Its Electric Scooters All Over the World," *The Verge*, August 1, 2018.

208. Jennifer Huddleston Skees and Trace Mitchell, "Will the Electric Scooter Movement Lose Its Charge?," Mercatus Center at George Mason University, *The Bridge*, July 18, 2018.

209. Tom Maguire, "A New Permit and Pilot Program for San Francisco's Scooters," San Francisco Municipal Transportation Agency, *Street Talk* (blog), May 1, 2018, https://www.sfmta.com/blog/new-permit-and-pilot-program-san-franciscos-scooters.

210. Ida Mojadad, "S.F. Will Be Scooter-Free While City Chooses Permit Holders," *SF Weekly*, May 24, 2018; James Brasuell, "Miami Sends Cease and Desist to Electric Scooter Companies," Planetizen, June 27, 2018.

211. Regina Clewlow, "The Micro-Mobility Revolution," *Medium*, July 24, 2018.

212. Kevin Roose, "How I Learned to Stop Worrying and Love Electric Scooters," *New York Times*, June 6, 2018; Nick Zaiac, "Not a Fad: Why Shared Bikes, Scooters Beat Taking the Bus," American Institute for Economic Research, June 19, 2018.

213. Skees and Mitchell, "Will the Electric Scooter Movement Lose Its Charge?"

214. Alex Hern, "The Future Will Be Dockless: Could a City Really Run on 'Floating Transport?,'" *The Guardian*, July 11, 2018.

215. Citymapper, "The Age of Floating Transport," *Medium*, July 2, 2018.

216. Hern, "The Future Will Be Dockless."

217. Adam Thierer and Caleb Watney, "Robots Don't Get Drunk or Drowsy, So Why Hold Up Driverless Cars?," *The Hill*, July 14, 2017.

218. Thierer and Watney, "Robots Don't Get Drunk."

219. Elon Musk (@elonmusk), "Some Exciting News This Week: Tesla Version 7 Software with Autopilot Goes to Wide Release on Thursday!," Twitter, October 10, 2015, 7:35 p.m., https://twitter.com/elonmusk/status/653005894922276865.

220. Tesla, "Your Autopilot Has Arrived," October 14, 2015.

221. NHTSA, "Regulations," accessed May 24, 2018, https://www.nhtsa.gov/laws-regulations/fmvss.

222. Andrew J. Hawkins, "The Federal Government Doesn't Know How to Regulate Tesla's Autopilot Software," *The Verge*, November 4, 2015.

223. "Federal Automated Vehicle Policy: Accelerating the Next Revolution in Roadway Safety," National Highway Traffic Safety Administration, U.S. Department of Transportation, September 2016, p. 76.

224. "Federal Automated Vehicle Policy," pp. 76–77.

225. Adam Thierer and Caleb Watney, Comment on the Federal Automated Vehicles Policy, Mercatus Center at George Mason University, *Public Interest Comment*, December 5, 2016.

226. "Preparing for the Future of Transportation: Automated Vehicle 3.0," NHTSA, U.S. Department of Transportation, December 22, 2018; Jennifer Huddleston and Trace Mitchell, "Continuing DOT's Automated Vehicle Soft-Law Approach Will Encourage Innovation and Promote Safety," Mercatus Center at George Mason University, *Public Interest Comment*, November 30, 2018.

227. Jane Komsky, "Addressing the Dangers of Partially Driverless Cars," *The Regulatory Review*, February 6, 2018; Tracy Pearl, "Hands on the Wheel: A Call for Greater Regulation of Semi-Autonomous Cars," *Indiana Law Journal*, March 8, 2017.

228. Andrew Askland, "Introduction: Why Law and Ethics Need to Keep Pace with Emerging Technologies," in *The Growing Gap Between Emerging Technologies and Legal-Ethical Oversight: The Pacing Problem*, ed. Gary E. Marchant, Braden R. Allenby, and Joseph R. Herkert (Dordrecht, Neth.: Springer, 2011), p. xviii.

229. Alex Heath, "Meet Geohot, the Guy Who Unlocked the First iPhone and Hacked the Sony PS3," Cult of Mac, April 30, 2012.

230. National Highway Traffic Safety Administration, Special Order Directed to Comma.ai, October 27, 2016.

231. Comma (@comma_ai), "Got this in the mail today. https://www.scribd.com/document/329218929/2016-10-27-Special-Order-Directed-to-Comma-ai . . . First time I hear from them and they open with threats. No attempt at a dialog. -GH 1/3," Twitter, October 28, 2016, 4:02 a.m., https://twitter.com/comma_ai/status/791958356042719234.

232. Comma (@comma_ai), "Would much rather spend my life building amazing tech than dealing with regulators and lawyers. It isn't worth it. -GH 2/3," Twitter, October 28, 2016, 4:02 a.m., https://twitter.com/comma_ai/status/791958385348321284.

233. Comma (@comma_ai), "The comma one is cancelled. comma.ai will be exploring other products and markets. Hello from Shenzhen, China. -GH 3/3," Twitter, October 28, 2016, 4:02 a.m., https://twitter.com/comma_ai/status/791958413345382400.

234. Brad Templeton, "Comma.ai Cancels Comma-One Add-On Box After Threats from NHTSA," *Robohub*, October 31, 2016; Kyle Stock, "NHTSA Scared This Self-Driving Entrepreneur off the Road," Bloomberg Tech, October 28, 2016.

235. Megan Geuss, "After Mothballing Comma One, George Hotz Releases Free Autonomous Car Software," Ars Technica, November 30, 2016.

236. Alex Davies, "Americans Can't Have Audi's Super Capable Self-Driving System," *Wired*, May 15, 2018.

237. Jerry Brito and Andrea Castillo, *Bitcoin: A Primer for Policymakers* (Arlington, VA: Mercatus Center at George Mason University, August 19, 2013).

238. Satoshi Nakamoto, "Bitcoin Open Source Implementation of P2P Currency," P2P Foundation, February 11, 2009.

239. Nakamoto, "Bitcoin Open Source Implementation."

240. Coindesk, "What Is Blockchain Technology?," accessed October 31, 2018.

241. Larry White, "Bitcoin After 10 Years," *Alt-M* (blog), October 23, 2018.

242. Internal Revenue Service, Internal Revenue Bulletin: 2014–16, April 14, 2014.

243. Evelyn Cheng, "Hardly Anyone Is Paying Taxes on Their Bitcoin Gains as Filing Deadline Nears," CNBC, April 13, 2018.

244. Andrea O'Sullivan, "Private Money in Virtual Worlds," *Reason Magazine*, March 2015.

245. Diego Zuluaga, "Of Libras and Zebras, Part Three," *Alt-M* (blog), July 17, 2019; Andrea O'Sullivan, "Congressional Libra Hearings Show It's Tough to Innovate on Compliance," Mercatus Center at George Mason University, *The Bridge*, July 18, 2019.

246. Edward Stringham and Max Gulker, "The Prospects of Decentralized Marketplaces with Privately Enforced Contracts: A Case Study of OpenBazaar," working paper, Mercatus Center at George Mason University, Arlington, VA, (forthcoming, 2020).

247. Jerry Brito, Houman Shadab, and Andrea Castillo, "Bitcoin Financial Regulation: Securities, Derivatives, Prediction Markets, and Gambling," *Columbia Science and Technology Law Review*, no. XVI (2014): 144–221.

248. Adam Ozimek, "The Regulation and Value of Prediction Markets," working paper, Mercatus Center at George Mason University, Arlington, VA, March 12, 2014.

249. For more on the theory and workings of information markets, see Robin Hanson, "Shall We Vote on Values, but Bet on Beliefs?," working paper, Mercatus Center at George Mason University, Arlington, VA, October 2007.

250. Jack Peterson et al., "Augur: A Decentralized Oracle and Prediction Market Platform," Augur, white paper, March 5, 2018.

251. Peter van Valkenburgh, "Framework for Securities Regulation of Cryptocurrencies," Coin Center Policy Report, August 10, 2018, https://coincenter.org/entry/framework-for-securities-regulation-of-cryptocurrencies.

252. One example is SatoshiDice, where players can participate in a simple dice game on the Bitcoin blockchain. Because of legal headaches, SatoshiDice operators decided to ban any activity originating from IP addresses in the United States.

253. Christopher Koopman, "The Doomed Crusade Against Daily Fantasy Sports," *National Review*, November 16, 2015.

254. Adam Silver, "Legalize and Regulate Sports Betting," *New York Times*, November 13, 2014.

255. Christopher Koopman and Michael Kotrous, "It's a New Day for Daily Fantasy Sports Betting," *The Hill*, June 5, 2018.

256. Tyler Cowen, "Don't Let Doubts About Blockchains Close Your Mind," Bloomberg View, April 27, 2018.

257. Cowen, "Don't Let Doubts About Blockchains Close Your Mind."

258. Jonathan Logan, "Dropgangs, or the Future of Dark Markets," *Opaque Link* (blog), December 26, 2018.

259. Cory Doctorow, "Dark Markets Have Evolved to Use Encrypted Messengers and Dead-Drops," *Boing Boing*, (blog), January 14, 2019.

260. Andrea O'Sullivan, "'Drop Gangs:' The Latest Evolution in Darknet's Avoidance of Law Enforcement," *Reason*, January 22, 2019.

261. Adam Thierer, "Innovation Arbitrage, Technological Civil Disobedience & Spontaneous Deregulation," *Medium*, December 7, 2016.

262. Benjamin Edelman and Damien Geradin, "Spontaneous Deregulation," *Harvard Business Review*, April 2016.

263. Adam Thierer, "What 20 Years of Internet Law Teaches Us about Innovation Policy," *Federalist Society* (blog), May 12, 2016.

264. Brian R. Knight, "Modernizing Regulation to Encourage Fintech Innovation: Examining Opportunities and Challenges in the Financial Technology (Fintech) Marketplace," Mercatus Center at George Mason University, Congressional testimony before the House Committee on Financial Services, Subcommittee on Financial Institutions and Consumer Credit, 115th Cong., 2nd sess., January 30, 2018, p. 4.

265. Brian Knight, "Federalism and Federalization on the Fintech Frontier," *Vanderbilt Journal of Entertainment and Technology Law* 20, no. 1 (2017).

266. Eli Dourado, "The FAA Is Constantly Thwarting Innovation," *Slate*, February 17, 2016; Eli Dourado and Michael Kotrous, "Airplane Speeds Have Stagnated for 40 Years," *Mercatus Center Chart*, July 20, 2016.

267. Lynne Kiesling and Mark Silberg, "Regulation, Innovation, and Experimentation: The Case of Residential Rooftop Solar," *Annual Proceedings of the Wealth and Well-Being of Nations*, vol. VI, Beloit College, 2015, pp. 99–126.

Chapter 3

1. A. John Simmons, "The Duty to Obey and Our Natural Moral Duties," in *Is There a Duty to Obey the Law?*, Christopher Heath Wellman and A. John Simmons (Cambridge: Cambridge University Press, 2005), p. 96.

2. Tom R. Tyler, *Why People Obey the Law* (New Haven, CT: Yale University Press, 1990).

3. Jonathan Mayhew, *A Discourse Concerning Unlimited Submission and Non-Resistance to the Higher Powers*, (1750), DigitalCommons@ University of Nebraska–Lincoln, *Electronic Texts in American Studies* 44, p. 29.

4. Mayhew, *A Discourse.*

5. Gordon Wood, *The Radicalism of the American Revolution* (New York: Vintage Books, 1991).

6. Wood, *The Radicalism of the American Revolution*, p. 17.

7. Henry David Thoreau, "Civil Disobedience" 1849, http://utc.iath .virginia.edu/abolitn/abeshdtat.html

8. Rev. Martin Luther King Jr., "Letter from a Birmingham Jail," April 16, 1963, https://kinginstitute.stanford.edu/encyclopedia/letter-birmingham-jail.

9. Theresa Züger, "Re-thinking Civil Disobedience," *Internet Policy Review* 2, no. 4, (2013).

10. Howard Zinn, "The Problem Is Civil Obedience," in *Violence: The Crisis of American Confidence*, ed. Hugh Davis Graham (Baltimore: Johns Hopkins Press, 1972).

11. Karla Adams, "Occupy Wall Street Protests, Movement Continues to Spread One Month Later," *Washington Post*, October 17, 2011.

12. Finn Brunton and Helen Nissenbaum, *Obfuscation: A User's Guide for Privacy and Protest* (Cambridge, MA: MIT Press, 2015), p. 1.

13. Étienne de La Boétie, "The Discourse of Voluntary Servitude," (1576), trans. Harry Kurz, *Online Library of Liberty.*

14. Charles Murray, *By the People: Rebuilding Liberty without Permission* (New York: Crown Forum, 2016), p. 119.

15. Timothy C. May, "The Crypto Anarchist Manifesto," November 22, 1992, https://www.activism.net/cypherpunk/crypto-anarchy.html.

16. John Perry Barlow, *A Declaration of the Independence of Cyberspace*, Electronic Frontier Foundation, Davos, Switz., Feb. 8, 1996.

17. John Rawls, *A Theory of Justice* (Cambridge, MA: Harvard University Press, 1971), p. 368.

18. Hannah Arendt, *Crises of the Republic: Lying in Politics, Civil Disobedience, On Violence, Thoughts on Politics and Revolution* (Orlando, FL: Harcourt Brace, 1972), p. 74.

19. Alexis de Tocqueville, "Political Associations in the United States," in *Democracy in America*, ed. Bruce Frohnen (Washington: Regnery Publishing, Inc., 2002), p. 151.

20. Thomas Jefferson, Letter to James Madison, January 30, 1787, National Archives, archives.gov.

21. Thomas Jefferson, Letter to William Stephens Smith, November 13, 1787, in *The Works of Thomas Jefferson*, Federal Edition, vol. 5, ed. Paul Leicester Ford (New York and London: G. P. Putnam's Sons, 1904–05), viewed on *Online Library of Liberty*.

22. Arendt, *Crises of the Republic*, p. 102.

23. Arendt, *Crises of the Republic*, p. 102.

24. Emanuela Carbonara, Francesco Parisi, and Georg von Wangenheim, "Legal Innovation and the Compliance Paradox," *Minnesota Journal of Law, Science & Technology* 9, no. 2 (2008): 842.

25. Carbonara, Parisi, and von Wangenheim, "Legal Innovation and the Compliance Paradox," 842.

26. Carbonara, Parisi, and von Wangenheim, "Legal Innovation and the Compliance Paradox," 853. See also William J. Stuntz, "Self-Defeating Crimes," *Virginia Law Review* 86, no. 8, Symposium: The Legal Construction of Norms, November 2000, p. 1875 ("Police and prosecutors have to decide where to invest their time and energy, what illegal markets to attack, because they cannot possibly attack all of them.").

27. Arendt, *Crises of the Republic*, p. 74.

28. Pew Research Center, "Public Trust in Government: 1958–2017," December 14, 2017.

29. Peter A. French, "Foreword," in *The Growing Gap Between Emerging Technologies and Legal-Ethical Oversight: The Pacing Problem*, ed. Gary E. Marchant, Braden R. Allenby, and Joseph R. Herkert (Dordrecht, Neth.: Springer, 2011), p. vi.

30. Elizabeth Pollman, "Corporate Disobedience," *Duke Law Journal* 68 (2019): 709.

31. Niklas Elert and Magnus Henrekson, "Evasive Entrepreneurialism," *Small Business Economics* 47 (2016): 106. See also Michael Farren, "Ending the Uber Wars: How to Solve a Special Interest Nightmare," *Fiscal Times*, Aug. 11, 2015.

32. Pollman, "Corporate Disobedience": 731.

33. Christopher Coyne and Peter Boettke, "An Entrepreneurial Theory of Social and Cultural Change," in *Markets and Civil Society: The European Experience in Comparative Perspective*, ed. Víctor Pérez-Díaz (New York: Berghahn Books, 2009), pp. 77–103.

34. Food and Drug Administration, "Information about Self-Administration of Gene Therapy," page last updated 11/21/2017, https://www.fda.gov/BiologicsBloodVaccines/CellularGeneTherapyProducts/ucm586343.htm.

35. Jason F. Brennan and Peter Jaworski, *Markets without Limits: Moral Virtues and Commercial Interests* (New York: Routledge, 2016), p. 17.

36. Timothy Sandefur, *The Right to Earn a Living: Economic Freedom and the Law* (Washington: Cato Institute, 2010).

37. David S. Lucas and Caleb S. Fuller, "Entrepreneurship: Productive, Unproductive, and Destructive—Relative to What?," *Journal of Business Venturing Insights* 7 (2015): 47.

38. Lucas and Fuller, "Entrepreneurship."

39. Elert and Henrekson, "Evasive Entrepreneurialism," p. 32.

40. Joel Feinberg, "Civil Disobedience in the Modern World," in *Humanities in Society* 2, no. 1 (1979): 37.

41. Tom Angell, "Sessions Admits Feds Can't Effectively Police Marijuana in States," Marijuana Moment, March 10, 2018.

42. Rachana Pradhan, "Vermont Becomes First State to Permit Drug Imports from Canada," *Politico*, May 16, 2018.

43. Janell Ross, "6 Big Things to Know about Sanctuary Cities," *Washington Post*, July 8, 2015.

44. Jacob Gershman and Dan Frosch, "Rural Sheriffs Defy New Gun Measures," *Wall Street Journal*, March 10, 2019.

45. Melanie Ehrenkranz, "Basically No One Is Getting Fined for Flying Drones without a License," Gizmodo, February 9, 2018.

46. Sally French, "Exclusive: Only One Drone Pilot Has Ever Been Busted for Flying without a License—and He Got a Warning," *MarketWatch*, February 19, 2018.

47. Ehrenkranz, "Basically No One."

48. Carbonara, Parisi, and von Wangenheim, "Legal Innovation and the Compliance Paradox," p. 853.

49. Federal Aviation Administration, "Law Enforcement Guidance for Suspected Unauthorized UAS Operations," Version 4, June 5, 2017.

50. Shaun Courtney, "Chao Defying Law on Self-Driving Auto Test Funds, Lawmakers Say," Bloomberg Government, November 5, 2018.

51. Ryan Hagemann, Jennifer Huddleston Skees, and Adam Thierer, "'Soft Law' Is Eating the World: Driverless Car Edition," Mercatus Center at George Mason University, *The Bridge*, October 11, 2018.

52. John C. Carey, "Is Rulemaking Old Medicine at the FDA?," 1997 Third Year Paper, Harvard Law School, Digital Access to Scholarship at Harvard, *HLS Student Papers* (1997).

53. Food and Drug Administration, *Selected FDA GCP/Clinical Trial Guidance Documents*, September 25, 2017.

54. Food and Drug Administration, *Mobile Medical Applications: Guidance for Industry and Food and Drug Administration Staff*, February 9, 2015.

55. Food and Drug Administration, *Technical Considerations for Additive Manufactured Medical Devices: Guidance for Industry and Food and Drug Administration Staff*, December 5, 2017.

56. Philip J. Weiser, "Entrepreneurial Administration," *Boston University Law Review* 97 (2017).

Chapter 4

1. Richard A. Epstein, "Why the Modern Administrative State Is Inconsistent with the Rule of Law," *NYU Journal of Law & Liberty* 3 (2008): 491–515.

2. Paul Kane and Derek Willis, "Laws and Disorder," *Washington Post*, November 5, 2018.

3. Larry Downes, "The Right and Wrong Ways to Regulate Self-Driving Cars," *Harvard Business Review*, December 6, 2016.

4. Janel Davis, "Your Permit, Please?," *Memphis Flyer*, September 30, 2004.

5. Paul Mulshine, "Nutty New Jersey Laws: Do You Honk Your Horn When Passing? You're Required To," *The Star Ledger*, March 3, 2012.

6. Mulshine, "Nutty New Jersey Laws."

7. Louis Anslow, "Headphone Hysteria and the Man Who Broke the Law for Walkman Rights," *Medium*, September 2016.

8. Anslow, "Headphone Hysteria."

9. Carly Baldwin, "Walkman Banned in Woodbridge? Yes, Law Is Still on the Books," Woodbridge Patch, September 30, 2016.

10. Timothy B. Lee, "Stealing 'Entertainment Services' Now a Crime in Tennessee," Ars Technica, June 2, 2011.

11. "Florida Accidentally Banned All Computers, Smart Phones in the State through Internet Cafe Ban: Lawsuit," HuffPost, July 9, 2012.

12. Alex Davies, "Americans Can't Have Audi's Super Capable Self-Driving System," *Wired*, May 15, 2018. ("New York's 1971 requirement that drivers keep at least one hand on the steering wheel could cause trouble, should anyone decide to enforce it. California's new regulations for commercial deployment of robo-cars might subject Audi to a bevy of requirements, from customer education, including for those who buy the car used, to data collection in case of a collision.")

13. Sally French, "Exclusive: Only One Drone Pilot Has Ever Been Busted for Flying without a License—and He Got a Warning," *MarketWatch*, February 19, 2018.

14. Philip K. Howard, "Radically Simplify Law," Cato Institute, Cato Online Forum, http://www.cato.org/publications/cato-online-forum/radically-simplify-law.

15. Clyde Wayne Crews Jr., *Ten Thousand Commandments: An Annual Snapshot of the Federal Regulatory State* (Washington: Competitive Enterprise Institute, 2018).

16. Crews Jr., *Ten Thousand Commandments*, p. 4.

17. Crews Jr., *Ten Thousand Commandments*, p. 3.

18. Crews Jr., *Ten Thousand Commandments*, p. 3.

19. Crews Jr., *Ten Thousand Commandments*, p. 3.

20. James Broughel, *Regulation and Economic Growth: Applying Economic Theory to Public Policy* (Mercatus Center at George Mason University, 2017); James Broughel, "Why Do We Get So Much Regulation?," *Neighborhood Effects* (blog), Mercatus Center at George Mason University, February 17, 2017.

21. Patrick McLaughlin and Michael Wilt, "Regulatory Accumulation: The Problem and Solutions," *Policy Spotlight*, Mercatus Center at George Mason University, September 2017.

22. McLaughlin and Wilt, "Regulatory Accumulation."

23. Daniel Byler, Beth Flores, and Jason Lewris, "Using Advanced Analytics to Drive Regulatory Reform: Understanding Presidential Orders on Regulation Reform," Deloitte, 2017.

24. Bentley Coffey, Patrick McLaughlin, and Pietro Peretto, "The Cumulative Cost of Regulations," working paper, Mercatus Center at George Mason University, Arlington, VA, April 2016.

25. Clyde Wayne Crews Jr., "Liberty's Unfinished Business: How to Eliminate Political Barriers to Global Entrepreneurship," in *Demographics and Entrepreneurship: Mitigating the Effects of an Aging Population*, ed. Steven Globerman and Jason Clemons (Canada: Fraser Institute, 2018), p. 293.

26. John Malcolm, "Criminal Law and the Administrative State: The Problem with Criminal Regulations," Heritage Foundation, *Legal Memorandum*, no. 130, August 6, 2014.

27. "Overcriminalization," *Solutions 2018: The Policy Briefing Book*, Heritage Foundation, 2018.

28. "Overcriminalization," *Solutions 2018*.

29. Adam Thierer and Jennifer Huddleston Skees, "Lemonade Stands and Permits," Mercatus Center at George Mason University, *The Bridge*, August 20, 2018.

30. Jordan Reimschisel and Adam Thierer, "'Build & Freeze' Regulation Versus Iterative Innovation," *Plain Text*, November 1, 2017.

31. Cristie Ford, *Innovation and the State: Finance, Regulation, and Justice* (Cambridge: Cambridge University Press, 2017), p. 83.

32. Lyn M. Gaudet and Gary E. Marchant, "Administrative Law Tools for More Adaptive and Responsive Regulation," in *The Growing Gap Between Emerging Technologies and Legal-Ethical Oversight: The Pacing Problem*, ed. Gary E. Marchant, Braden R. Allenby, and Joseph R. Herkert (Dordrecht, Neth.: Springer, 2011), p. 167.

33. Daniel Castro and Alan McQuinn, "How and When Regulators Should Intervene," *Information Technology and Innovation Foundation Reports* (February 2015): 6.

34. James Q. Wilson, *Bureaucracy: What Government Agencies Do and Why They Do It* (New York: Basic Books, 1989), p. 69.

35. Henry I. Miller, "Follow the FDA's Self-Interest," *Wall Street Journal*, October 28, 2018.

36. Jonathan Adler, "The Problems with Precaution: A Principle without Principle," *The American*, May 25, 2011. See also Sam Kazman, "Deadly Overcaution: FDA's Drug Approval Process," *Journal of Regulation and Social Costs* (September 1990).

37. William K. Stevens, "Drug Approved to Prevent Ulcers in Arthritis Sufferers," *New York Times,* December 28, 1988.

38. Joseph V. Gulfo, *Innovation Breakdown: How the FDA and Wall Street Cripple Medical Advances* (Franklin, TN: Post Hill, 2014).

39. Marie Thibault, "FDA Approving Devices Faster," MD+DI, March 26, 2015.

40. Adam Thierer, *Permissionless Innovation: The Continuing Case for Comprehensive Technological Freedom*, 2nd ed. (Arlington, VA: Mercatus Center at George Mason University, 2016), pp. 51–56.

41. National Academies of Sciences, Engineering, and Medicine, *Assessing the Risks of Integrating Unmanned Aircraft Systems (UAS) into the National Airspace System: Consensus Study Report* (Washington: The National Academies Press, 2018), pp. 3–6.

42. Adam Thierer, "National Academies Report Rips FAA's Risk-Averse Regulatory Culture," *Technology Liberation Front* (blog), June 12, 2018.

43. Alan Levin, "Drones May Need License Plates Soon," *Bloomberg*, May 23, 2018.

44. Mark Febrizio and Patrick McLaughlin, "How Performance Standards Create Smarter Regulations Than Design Mandates," *SmartRegs* (blog), May 4, 2017.

45. Ruth B. Carter and Gary E. Marchant, "Principles-Based Regulation and Emerging Technology," in *The Growing Gap Between Emerging Technologies and Legal-Ethical Oversight: The Pacing Problem*, ed. Gary E. Marchant, Braden R. Allenby, and Joseph R. Herkert (Dordrecht, Neth.: Springer, 2011), pp. 157–66.

46. Niklas Elert and Magnus Henrekson, "Entrepreneurship and Institutions: A Bidirectional Relationship," Research Institute of Industrial Economics, IFN Working Paper no. 1153, 2017, p. 43.

47. Marc Scribner, "Outdated Auto Safety Regulations Threaten the Self-Driving Revolution," *Wired*, February 8, 2018.

48. Scribner, "Outdated Auto Safety Regulations."

49. Scribner, "Outdated Auto Safety Regulations."

50. Brent Skorup, "The Department of Transportation's Proposed Vehicle-to-Vehicle Technology Mandate Is Unprecedented and Hasty," Mercatus Center at George Mason University, *Public Interest Comment*, April 14, 2017.

51. Skorup, "Vehicle-to-Vehicle Technology," p. 2.

52. Eric von Hippel, *Free Innovation* (Cambridge, MA: MIT Press, 2017), p. 130.

53. Aaron Wildavsky, *Searching for Safety* (New Brunswick, CT: Transaction Books, 1988), p. 183.

54. James Bailey and Diana Thomas, "Regulating Away Competition: The Effect of Regulation on Entrepreneurship and Employment," working paper, Mercatus Center at George Mason University, Arlington, VA, September 2015.

55. Frédéric Bastiat, *What Is Seen and What Is Not Seen* (Indianapolis, IN: Liberty Fund, 1848, 1955); Per L. Bylund, *The Seen, the Unseen, and the Unrealized: How Regulations Affect Our Everyday Lives* (Lanham, MD: Lexington Books, 2016), pp. 73–83; Thierer, *Permissionless Innovation*, p. 13.

56. John Wu and Robert D. Atkinson, *How Technology-Based Start-Ups Support U.S. Economic Growth*, Information Technology and Innovation Foundation, November 28, 2017.

57. Joseph Schumpeter, *Capitalism, Socialism and Democracy*, 3rd ed., (New York: Harper Perennial, 1942, 2008), p. 84.

58. Tyler Cowen, "Lobbying Doesn't Help Companies or Their Shareholders," *Bloomberg Opinion*, January 11, 2018.

59. Kevin M. Murphy, Andrei Shleifer, and Robert W. Vishny, "Why Is Rent-Seeking So Costly to Growth?," *American Economic Review Papers and Proceedings* 83, no. 2 (1993): 413–14. See also Kevin M. Murphy, Andrei Shleifer, and Robert W. Vishny, "The Allocation of Talent: Implications for Growth," *Quarterly Journal of Economics* 106, no. 2 (1991): 503–30.

60. Jeff Rowes, "Caskets and the Constitution: How a Simple Box Has Advanced Economic Liberty," *Wake Forest Law Journal of Law & Policy* 8, no. 1 (2018): 65.

61. Mark Zachary Taylor, *The Politics of Innovation: Why Some Countries Are Better Than Others at Science and Technology* (Oxford: Oxford University Press, 2016), p. 213.

62. Brink Lindsey and Steven Teles, *The Captured Economy: How the Powerful Enrich Themselves, Slow Down Growth, and Increase Inequality* (Cambridge: Oxford University Press, 2018), p. 17.

63. Calestous Juma, *Innovation and Its Enemies: Why People Resist New Technologies* (New York: Oxford University Press, 2016).

64. Alan Olmstead and Paul W. Rhode, "The Agricultural Mechanization Controversy of the Interwar Years," *Agricultural History* 68, no. 3 (Summer 1994): 36.

65. Matt Schruers, "Swampetition: Hobbling Rivals with Regulation, from Horses to the Internet," *DisCo* (Disruptive Competition Project), May 17, 2018.

66. David Pogue, "Why You Can't Buy a Tesla in These States," *Yahoo Finance*, October 15, 2018.

67. Jerry Ellig and Jesse Martinez, "Auto Franchise Laws Restrict Consumer Choice and Increase Prices," Mercatus Center at George Mason University, March 30, 2015.

68. Daniel O'Connor, "Tesla, the Auto Dealers and New Jersey: Playing the Consumer Protection Card," *DisCo* (Disruptive Competition Project), March 11, 2014.

69. Robert D. Atkinson, "The 2014 ITIF Luddite Awards," Information Technology and Innovation Foundation, January 5, 2015; Information Technology and Innovation Foundation, "Efforts to Block Tesla Win ITIF's Luddite of the Year," February 6, 2015.

70. James Bailey and Diana Thomas, "Regulating Away Competition: The Effect of Regulation on Entrepreneurship and Employment," working paper, Mercatus Center at George Mason University, Arlington, VA, September 2015, p. 16.

71. Matthew D. Mitchell, *The Pathology of Privilege: The Economic Consequences of Government Favoritism* (Arlington, VA: Mercatus Center at George Mason University, 2012), pp. 1–2.

72. Jeffrey M. Berry, *The Interest Group Society*, 2nd ed. (Glenview, IL: Scott, Foresman & Co., 1989).

73. Theodore J. Lowi, *The End of Liberalism: The Second Republic of the United States*, 2nd ed. (New York: W. W. Norton, 1969, 1979).

74. James Q. Wilson said that client politics "occurs when most or all of the benefits of a program go to some single, reasonably small interest (and industry, profession, or locality) but most or all of the costs will be borne by a large number of people (for example, all taxpayers)." James Q. Wilson, *Bureaucracy* (New York: Basic Books, 1989), p. 76.

75. Mancur Olson, *The Logic of Collective Action: Public Goods and the Theory of Groups* (Cambridge, MA: Harvard University Press, 1965).

76. A. John Simmons, "The Duty to Obey and Our Natural Moral Duties," in *Is There a Duty to Obey the Law?*, Christopher Heath Wellman and A. John Simmons (Cambridge: Cambridge University Press, 2005), p. 64.

77. Michael Giberson, "Concentrated Benefits and Dispersed Costs," *Knowledge Problem* (blog), October 17, 2010.

78. Christopher Coyne and Peter Boettke, "An Entrepreneurial Theory of Social and Cultural Change," in *Markets and Civil Society: The European Experience in Comparative Perspective*, ed. Víctor Pérez-Díaz (New York: Berghahn Books, 2009), p. 181.

79. Adam Thierer, *Media Myths: Making Sense of the Debate over Media Ownership* (Washington: Progress & Freedom Foundation, 2005).

80. Brent Skorup, "Who Needs the FCC?," *National Affairs* 36 (Winter 2016).

81. Jesse Walker, *Rebels on the Air: An Alternative History of Radio in America* (New York: NYU Press, 2001).

82. David Weinstein, *The Forgotten Network: DuMont and the Birth of American Television* (Philadelphia: Temple University Press, 2004).

83. Jonathan Emord, *Freedom Technology and the First Amendment* (San Francisco: Pacific Research Institute, 1991), p. 265.

84. Adam Thierer, "Unnatural Monopoly: Critical Moments in the Development of the Bell System Monopoly," *Cato Journal* 14, no. 2 (Fall 1994): 267–85.

85. Adam Thierer and Brent Skorup, "A History of Cronyism and Capture in the Information Technology Sector," *Journal of Technology Law & Policy* 18 (2013).

86. Morris M. Kleiner and Evgeny S. Vorotnikov, *At What Cost? State and National Estimates of the Economic Costs of Occupational Licensing* (Arlington, VA: Institute for Justice, 2018), p. 5.

87. Patrick A. McLaughlin, Matthew D. Mitchell, and Anne Philpot, "The Effects of Occupational Licensure on Competition, Consumers, and the Workforce," Mercatus Center at George Mason University, *Public Interest Comment*, November 2017.

88. U.S. Department of the Treasury, Council of Economic Advisers, and U.S. Department of Labor, *Occupational Licensing: A Framework for Policymakers*, July 2015, p. 4.

89. U.S. Department of the Treasury, Council of Economic Advisers, and U.S. Department of Labor, *Occupational Licensing*.

90. Morris M. Kleiner, "Reforming Occupational Licensing Policies," Brookings Institution, The Hamilton Project, Discussion Paper 2015–01, January 2015, p. 6. See also Morris M. Kleiner and Alan B. Krueger, "The Prevalence and Effects of Occupational Licensing," National Bureau of Economic Research (NBER) Working Paper no. 14308, Cambridge, MA, September 2008.

91. U.S. Department of the Treasury, Council of Economic Advisers, and U.S. Department of Labor, *Occupational Licensing*, p. 13. See also Kleiner, *Reforming Occupational Licensing Policies*, p. 12.

92. U.S. Department of the Treasury, Council of Economic Advisers, and U.S. Department of Labor, *Occupational Licensing*, pp. 4–5.

93. Veronique de Rugy, "Occupational Licensing: Bad for Competition, Bad for Low-Income Workers," Mercatus Center at George Mason University, March 25, 2014; Kleiner, *Reforming Occupational Licensing Policies*, pp. 13–14.

94. Matthew D. Mitchell, "Occupational Licensing and the Poor and Disadvantaged," *Policy Spotlight*, Mercatus Center at George Mason University, September 28, 2017; Archbridge Institute, "Too Much License? A Closer Look at Occupational Licensing and Economic Mobility," *Inquiry & Analysis*, April 2018; Alex Muresianu, "Occupational Licensing Reform: A Bipartisan Blueprint for Helping Low-Income Workers," Foundation for Economic Education, FEE.org, July 13, 2018.

95. John Blevins, "License to Uber: Using Administrative Law to Fix Occupational Licensing," *UCLA Law Review* 64 (2017): 869.

96. Kleiner, *Reforming Occupational Licensing Policies*, p. 12. See also Brief of Antitrust Scholars as Amici Curiae in Support of Respondent, *North Carolina State Board of Dental Examiners v. FTC* (August 6, 2014), p. 2.

97. Blevins, "License to Uber": 869. See also Robert J. Thornton and Edward J. Timmons, "The De-licensing of Occupations in the United States," U.S. Bureau of Labor Statistics, *Monthly Labor Review*, May 2015, p. 13.

98. Thornton and Timmons, "The De-licensing of Occupations in the United States."

99. Adam Thierer, "Regulatory Capture: What the Experts Have Found," *Technology Liberation Front* (blog), December 19, 2010.

100. Alfred E. Kahn, *The Economics of Regulation: Principles and Institutions*, vol. 2 (Cambridge, MA: MIT Press, 1971), pp. 12, 46.

101. Philip J. Weiser, "Alfred Kahn as a Case Study of a Political Entrepreneur: An Essay in Honor of His 90th Birthday," *Review of Network Economics* 7, no. 4 (2008).

102. Matthew D. Mitchell, *The Pathology of Privilege: The Economic Consequences of Government Favoritism* (Arlington, VA: Mercatus Center at George Mason University, July 8, 2012).

103. Philip K. Howard, *The Death of Common Sense: How Law Is Suffocating America* (New York: Random House, 1994, 2011), p. 202.

104. *Powers v. Harris*, 379 F.3d 1208, 1221 (10th Circuit 2004).

105. Mark Zachary Taylor, *The Politics of Innovation*.

106. Hannah Arendt, *Crises of the Republic: Lying in Politics, Civil Disobedience, On Violence, Thoughts on Politics and Revolution* (Orlando, FL: Harcourt Brace, 1972), p. 89.

107. Peter J. Wallison, "Judges Can Check the Administrative State," *Wall Street Journal*, April 5, 2018.

108. Lowi, *The End of Liberalism*.

109. Jonathan Rauch, *Government's End: Why Washington Stopped Working* (1999), p. 125.

110. Rauch, *Government's End*, p. 152.

111. Steven Teles, "Kludgeocracy in America," *National Affairs* (Fall 2013), https://www.nationalaffairs.com/publications/detail/kludgeocracy -in-america.

112. Teles, "Kludgeocracy in America."

113. Jeff Stein, "A Staff Survey Shows Just How Broken Congress Is," *Vox*, August 8, 2017.

114. Congressional Management Foundation (CMF), *State of the Congress: Staff Perspectives on Institutional Capacity in the House and Senate* (Washington: CMF, 2017), p. 9.

115. Avi Selk, "'There's So Many Different Things!': How Technology Baffled an Elderly Congress in 2018," *Washington Post*, January 2, 2019.

116. Teles, "Kludgeocracy in America."

117. Richard Epstein, *Simple Rules for a Complex World* (Cambridge, MA: Harvard University Press, 1995).

118. Carter and Marchant, "Principles-Based Regulation and Emerging Technology," p. 157.

119. Kristina Nyström, "Business Regulation and Red Tape in the Entrepreneurial Economy," *CESIS Electronic Working Paper Series*, Working Paper no. 225 (Centre of Excellence for Science and Innovation Studies, March 2010), p. 2.

120. Adam Thierer, "The Pacing Problem, the Collingridge Dilemma & Technological Determinism," *Technology Liberation Front* (blog), August 16, 2018.

121. Wendell Wallach, *A Dangerous Master: How to Keep Technology from Slipping Beyond Our Control* (New York: Basic Books, 2015), p. 251.

122. Gary E. Marchant, "The Growing Gap Between Emerging Technologies and the Law," in *The Growing Gap Between Emerging Technologies and Legal-Ethical Oversight: The Pacing Problem*, ed. Gary E. Marchant, Braden R. Allenby, and Joseph R. Herkert (Dordrecht, Neth.: Springer, 2011), p. 19. See also Richard A. Posner, *Catastrophe: Risk and Response* (Oxford: Oxford University Press, 2004), p. 70.

123. Wallach, *A Dangerous Master*, p. 60.

124. Larry Downes, *The Laws of Disruption: Harnessing the New Forces That Govern Life and Business in the Digital Age* (New York: Basic Books, 2009), p. 2.

125. Remarks of FAA Administrator Michael Huerta, South by Southwest press event, FAA, Austin, Texas, March 14, 2016.

126. U.S. Department of Transportation, *Federal Automated Vehicles Policy* (September 2016), p. 8; U.S. Department of Transportation, *Preparing for the Future of Transportation: Automated Vehicles 3.0* (October 2018), p. 7.

127. Denise Grady and Sheila Kaplan, "F.D.A. Moves to Stop Rogue Clinics from Using Unapproved Stem Cell Therapies," *New York Times*, May 9, 2018.

128. National Human Genome Research Institute, "DNA Sequencing Costs: Data," accessed January 24, 2019, https://www.genome.gov/27541954/dna-sequencing-costs-data/.

129. Marchant, "The Growing Gap Between Emerging Technologies and the Law."

130. von Hippel, *Free Innovation*, p. 13.

131. Ford, *Innovation and the State*, p. 167.

132. Marchant, "The Growing Gap Between Emerging Technologies and the Law," p. 27.

133. Albert H. Segars, "Seven Technologies Remaking the World," *MIT Sloan Management Review*, March 9, 2018.

134. Marc Andreessen, "Why Software Is Eating the World," *Wall Street Journal*, August 20, 2011.

135. Ryan Hagemann, "New Rules for New Frontiers: Regulating Emerging Technologies in an Era of Soft Law," *Washburn Law Journal* 57, no. 2 (Spring 2018): 263.

136. Hagemann, "New Rules for New Frontiers."

Chapter 5

1. Steven Teles, "Kludgeocracy in America," *National Affairs* (Fall 2013), https://www.nationalaffairs.com/publications/detail/kludgeocracy-in-america.

2. Adam Thierer, *The Delicate Balance: Federalism, Interstate Commerce, and Economic Freedom in the Technological Age* (The Heritage Foundation, 1999).

3. Matthew D. Mitchell. "That Government Is Best Which Is Not Captured by Special Interests," *Capitol Hill, State House, or City Hall: Debating the Location of Political Power and Decision-Making*, Mercatus Center at George Mason University, July 21, 2017.

4. Teles, "Kludgeocracy in America."

5. Christopher DeMuth, "Can the Administrative State Be Tamed?," *Journal of Legal Analysis* 8, no. 1 (June 2016): 121–90; Peter J. Wallison, "Judges Can Check the Administrative State," *Wall Street Journal*, April 5, 2018.

6. Elizabeth H. Slattery, "Who Will Regulate the Regulators? Administrative Agencies, the Separation of Powers, and Chevron Deference," Heritage Foundation, *Legal Memorandum*, May 7, 2015.

7. Tyler Cowen, "Does Technology Drive the Growth of Government?," unpublished manuscript, June 22, 2009.

8. Cowen, "Does Technology Drive the Growth of Government?"

9. Jason Kuznicki, *Technology and the End of Authority: What Is Government For?* (Cham, Switz.: Palgrave Macmillan, 2017), p. 240.

10. Samuel C. Florman, *The Existential Pleasures of Engineering*, 2nd ed. (New York: St. Martin's Griffin, 1994), p. 63.

11. John Perry Barlow, *A Declaration of the Independence of Cyberspace*, Electronic Frontier Foundation, Davos, Switz., Feb. 8, 1996.

12. Barlow, *A Declaration of the Independence of Cyberspace.*

13. For a discussion of the early internet exceptionalist movement, see Will Rinehart, "Technologies of Freedom," Cato Institute, Cato Unbound, January 3, 2018, https://www.cato-unbound.org/2018/01/03/will-rinehart/technologies-freedom.

14. George Gilder, "Angst and Awe on the Internet," *Forbes ASAP*, December 4, 1995.

15. Nicholas Negroponte, *Being Digital* (New York: Vintage Books, 1995).

16. Thomas P. Hughes, *Human-Built World: How to Think about Technology and Culture* (Chicago: University of Chicago Press, 2004), p. 105.

17. Alex Kozinski and Josh Goldfoot, "A Declaration of the Dependence of Cyberspace," *Columbia Journal of Law & Arts* 32 (2009).

18. Jack Goldsmith and Tim Wu, *Who Controls the Internet: Illusions of a Borderless World* (Oxford: Oxford University Press, 2006).

19. Mireille Hildebrandt, *Smart Technologies and the End(s) of Law* (Northampton, MA: Edward Elgar, 2015).

20. Taylor Owen, *Disruptive Power: The Crisis of the State in the Digital Age* (Oxford: Oxford University Press, 2015), p. 42.

21. Robert Wright, *Nonzero: The Logic of Human Destiny* (New York: Vintage Books, 2001), p. 154.

22. Stephen J. Kobrin, "Back to the Future: Neomedievalism and the Postmodern Digital World Economy," in *Globalization and Governance*, ed. Aseem Prakash and Jeffrey Hart (London and New York: Routledge, 1999), p. 170.

23. Kobrin, "Back to the Future."

24. Rebecca MacKinnon, *Consent of the Networked: The Worldwide Struggle for Internet Freedom* (New York: Basic Books, 2012).

25. Adam Thierer and Clyde Wayne Crews Jr., *Who Rules the Net? Internet Governance and Jurisdiction* (Washington: Cato Institute, 2003).

26. David Post, "The 'Unsettled Paradox': The Internet, the State, and the Consent of the Governed," *Indiana Journal of Global Legal Studies* 5, no. 2 (Spring 1998): 526.

27. Adam Thierer, "Book Review: Consent of the Networked by Rebecca MacKinnon," *Technology Liberation Front* (blog), January 25, 2012.

28. Jenny Teichman and Katherine C. Evans, *Philosophy: A Beginner's Guide* (Cambridge, MA: Basil Blackwell, 1991), p. 105.

29. Adam Thierer, "Code, Pessimism, and the Illusion of 'Perfect Control,'" Cato Institute, Cato Unbound, May 8, 2009, https://www.cato-unbound.org/2009/05/08/adam-thierer/code-pessimism-illusion-perfect-control.

30. Clay Shirky, *Here Comes Everybody: The Power of Organizing without Organizations* (New York: Penguin Press, 2008).

31. Yochai Benkler, *The Wealth of Networks: How Social Production Transforms Markets and Freedom* (New Haven, CT: Yale University Press, 2006).

32. Martin Gurri, *The Revolt of the Public and the Crisis of Authority in the New Millennium* (San Francisco: Stripe Press, 2018).

33. Gurri, *The Revolt of the Public.*

34. Gurri, *The Revolt of the Public,* p. 75.

35. Gurri, *The Revolt of the Public,* p. 91.

36. Taylor Owen, *Disruptive Power,* pp. 10–13.

37. Gurri, *The Revolt of the Public.*

38. Richard Fontaine and Kara Frederick, "The Autocrat's New Tool Kit," *Wall Street Journal*, March 15, 2019.

39. Thomas Brewster, "The Feds Can Now (Probably) Unlock Every iPhone Model in Existence," *Forbes*, February 26, 2018.

40. Julia Angwin et al., "AT&T Helped U.S. Spy on Internet on a Vast Scale," *New York Times*, August 15, 2015.

41. Adrienne LaFrance, "The Convenience-Surveillance Tradeoff," *The Atlantic*, January 14, 2016.

42. Benjamin Wittes and Jodie C. Liu, "The Privacy Paradox: The Privacy Benefits of Privacy Threats," Center for Technology Innovation at Brookings, May 2015, p. 2.

43. Wittes and Liu, "The Privacy Paradox," p. 3.

44. Kuznicki, *Technology and the End of Authority*, p. 3.

45. Kuznicki, *Technology and the End of Authority*, p. 3.

46. Tyler Cowen, "Don't Let Doubts about Blockchains Close Your Mind," Bloomberg View, April 27, 2018.

47. Andrea O'Sullivan, "Ungoverned or Antigovernance? How Bitcoin Threatens the Future of Western Institutions," *Journal of International Affairs* 71, no. 2 (2018): 90–102.

48. Stephen J. Kobrin, "Back to the Future", p. 181.

49. Cristie Ford, *Innovation and the State: Finance, Regulation, and Justice* (Cambridge: Cambridge University Press, 2017), p. 16.

50. Ford, *Innovation and the State*, p. 17.

51. Albert O. Hirschman, *Exit, Voice, and Loyalty* (Cambridge, MA: Harvard University Press, 1970), p. 1.

52. Hirschman, *Exit, Voice, and Loyalty*, p. 82.

53. *Wall Street Journal* columnist Greg Ip has used the controlled or prescribed burn metaphor in relation to managing financial risk and other hazards. Greg Ip, *Foolproof: Why Safety Can Be Dangerous and How Danger Makes Us Safe* (New York: Little, Brown and Company, 2015), pp. 180–88.

54. Eli Dourado, "Technologies of Control and Resistance: Making Sense of Our Stagnant Dynamism," *Medium*, November 4, 2011.

55. Timothy Sandefur, *The Permission Society: How the Ruling Class Turns Our Freedoms into Privileges and What We Can Do About It* (New York: Encounter Books, 2016).

56. Andrea O'Sullivan, "'Drop Gangs': The Latest Evolution in Darknet's Avoidance of Law Enforcement," *Reason*, January 22, 2019.

57. Debora L. Spar, *Ruling the Waves: Cycles of Discovery, Chaos, and Wealth from the Compass to the Internet* (New York: Harcourt, 2001), pp. 7–8.

58. Vlad Tarko and Andrew Farrant, "The Efficiency of Regulatory Arbitrage," *Public Choice* 181 (2019): 141–66, https://doi.org/10.1007/s11127-018-00630-y.

59. Alfred C. Aman Jr., "Administrative Law for a New Century," in *The Province of Administrative Law*, ed. Michael Taggart (London: Hart Publishing, an imprint of Bloomsbury Publishing, 1997), p. 113.

60. Samuel Hammond, "Disrupting Bureaucracy," *Medium,* September 24, 2015.

61. Hammond, "Disrupting Bureaucracy."

62. Hirschman, *Exit, Voice, and Loyalty*, p. 117.

63. Paul Romer, "Interview on Urbanization, Charter Cities and Growth Theory," *Paul Romer* (blog), April 29, 2015.

Chapter 6

1. James Broughel and Adam Thierer, "Technological Innovation and Economic Growth: A Brief Report on the Evidence," Mercatus Center at George Mason University, research paper, February 2019.

2. Brett Frischmann, "Here's Why Tech Companies Abuse Our Data: Because We Let Them," *The Guardian*, April 10, 2018.

3. Kevin Roose, "Is Tech Too Easy to Use?," *New York Times*, December 12, 2018; William Powers, *Hamlet's BlackBerry: A Practical Philosophy for Building a Good Life in the Digital Age* (New York: HarperCollins, 2010).

4. Samuel C. Florman, *The Existential Pleasures of Engineering*, 2nd ed. (New York: St. Martin's Griffin, 1994), p. 77.

5. Emmanuel G. Mesthene, *Technological Change: Its Impact on Man and Society* (Cambridge, MA: Harvard University Press, 1970), pp. 47–48; Emmanuel G. Mesthene, "The Social Impact of Technological Change," in *Philosophy of Technology: The Technological Condition—An Anthology*, ed. Robert C. Scharff and Val Dusek (Malden, MA: Blackwell Publishing, 2003), p. 619.

6. Adam Thierer, "Muddling Through: How We Learn to Cope with Technological Change," *Medium*, June 30, 2014.

7. Brett Frischmann and Evan Selinger, *Re-Engineering Humanity* (Cambridge: Cambridge University Press, 2018).

8. This section adapted from: Adam Thierer, "Is It 'Techno-Chauvinist' & 'Anti-Humanist' to Believe in the Transformative Potential of Technology?," *Medium*, September 18, 2018; Adam Thierer, "The Pacing Problem, the Collingridge Dilemma & Technological Determinism," *Technology Liberation Front* (blog), August 16, 2018.

9. Sally Wyatt, "Technological Determinism Is Dead: Long Live Technological Determinism," in *The Handbook of Science and Technology Studies,* ed. Edward J. Hackett et al. (Cambridge, MA: MIT Press, 2008), p. 168.

10. Thomas P. Hughes, "Technological Momentum," in *Does Technology Drive History? The Dilemma of Technological Determinism*, ed. Merritt Roe Smith and Leo Marx (Cambridge, MA: MIT Press, 1994), p. 102. See also Ian Barbour, *Ethics in an Age of Technology* (New York: Harper One, 1993), p. xvii.

11. Leon Wieseltier, "Among the Disrupted," *New York Times*, January 7, 2015.

12. Siva Vaidhyanathan, "Techno-Fundamentalism Can't Save You, Mark Zuckerberg," *New Yorker*, April 21, 2018.

13. Evgeny Morozov, *To Save Everything, Click Here: The Folly of Technological Solutionism* (New York: Public Affairs, 2013).

14. Christopher Mims, "Driverless Hype Collides with Merciless Reality," *Wall Street Journal*, September 13, 2018. (Citing Meredith Broussard of New York University, who defines techno-chauvinism as "the idea that technology is always the highest and best solution, and is superior to the people-based solution.")

15. David Nordfors and Vint Cerf, "Make People Valuable Again," *TechCrunch*, November 4, 2018.

16. Wyatt, "Technological Determinism," p. 169.

17. Florman, *The Existential Pleasures*, pp. 48, 53.

18. Jacques Ellul, *La Technique ou L'enjeu du Siècle* (1954). Ellul's book was later translated into English: Jacques Ellul, *The Technological Society* (New York: Vintage Books, 1964). See also Doug Hill, "Jacques Ellul, Technology Doomsdayer before His Time," *Boston Globe*, July 8, 2015.

19. Lewis Mumford, *Technics and Civilization* (London: Routledge & Sons, 1934); Lewis Mumford, *Technics and Human Development* (New York: Harcourt Brace Jovanovich, 1966).

20. Neil Postman, *Technopoly: The Surrender of Culture to Technology* (New York: Vintage Books, 1992).

21. Postman, *Technopoly*, pp. 52, xii

22. Samuel C. Florman, *Blaming Technology: The Irrational Search for Scapegoats* (New York: St. Martin's Press, 1981), p. 22.

23. Aileen Graef, "Tech and Privacy Advocates Clash over Possibilities for Google Glass," PBS NewsHour Extra, August 9, 2013.

24. Alvin E. Roth, "Repugnance as a Constraint on Markets," *Journal of Economic Perspectives* 21, no. 3 (July 2007): 37–58.

25. Issie Lapowsky, "Americans Aren't Ready for the Future Google and Amazon Want to Build," *Wired*, April 14, 2014.

26. Adam Thierer, "The Great Facial Recognition Technopanic of 2019," Mercatus Center at George Mason University, *The Bridge*, May 17, 2019; Matthew Feeney, "Should Police Facial Recognition Be Banned?," Cato Institute, *Cato at Liberty* (blog), May 13, 2019, https://www.cato.org/blog/should-police-facial-recognition-be-banned.

27. Merritt Roe Smith, "Technological Determinism in American Culture," in *Does Technology Drive History? The Dilemma of Technological Determinism*, ed. Merritt Roe Smith and Leo Marx (Cambridge, MA: MIT Press, 1994): p. 2.

28. L. M. Sacasas, "Humanist Technology Criticism," *Technology, Culture, and Ethics* (blog), July 9, 2015.

29. Wieseltier, "Among the Disrupted."

30. Andrew McAfee, "Who Are the Humanists, and Why Do They Dislike Technology So Much?," *Financial Times*, July 7, 2015.

31. McAfee, "Who Are the Humanists."

32. Adam Thierer, *Permissionless Innovation: The Continuing Case for Comprehensive Technological Freedom*, 2nd ed. (Arlington, VA: Mercatus Center at George Mason University, 2016), pp. 33–38.

33. Richard B. Belzer, "Risk Assessment, Safety Assessment, and the Estimation of Regulatory Benefits," working paper, Mercatus Center at George Mason University, Arlington, VA, 2012, p. 5; John D. Graham and Jonathan Baert Wiener, eds., *Risk vs. Risk: Tradeoffs in Protecting Health and the Environment*, (Cambridge, MA: Harvard University Press, 1995).

34. Richard Posner, *Public Intellectuals: A Study of Decline* (Cambridge, MA: Harvard University Press, 2001), p. 285.

35. Stefan H. Thomke, *Experimentation Matters: Unlocking the Potential of New Technologies for Innovation* (Boston: Harvard Business Review Press, 2003), pp. 1, 4.

36. Florman, *The Existential Pleasures*, p. 58.

37. Florman, *The Existential Pleasures*, p. 6.

38. Norman Lewis, "The Enduring Wisdom of the Crowd," *Spiked*, October 24, 2018.

39. Adam Smith, *Theory of Moral Sentiments* (1759), book I, chap. 1, https://www.earlymoderntexts.com/assets/pdfs/smith1759.pdf.

40. Adam Thierer, "Adam Smith's Theory of Moral Sentiments Turns 250," *Technology Liberation Front* (blog), July 28, 2009.

41. Thomas L. Haskell, "Capitalism and the Origins of the Humanitarian Sensibility, Part 1," *American Historical Review* 90, no. 2 (April 1985): 356.

42. Haskell, "Capitalism and the Origins of the Humanitarian Sensibility."

43. Haskell, "Capitalism and the Origins of the Humanitarian Sensibility."

44. American Humanist Association, "Humanism and Its Aspirations: Humanist Manifesto III, a Successor to the Humanist Manifesto of 1933" (revised 2003).

45. Steven Pinker, *Enlightenment Now: The Case for Reason, Science, Humanism, and Progress* (New York: Viking, 2018), p. 410.

46. Adam Thierer, "Wendell Wallach on the Challenge of Engineering Better Technology Ethics," *Technology Liberation Front* (blog), April 20, 2016.

47. Wendell Wallach, *A Dangerous Master: How to Keep Technology from Slipping Beyond Our Control* (New York: Basic Books, 2015), pp. 247–66.

48. This section adapted from: Adam Thierer, "Are 'Permissionless Innovation' and 'Responsible Innovation' Compatible?," *Technology Liberation Front* (blog), July 12, 2017.

49. René Von Schomberg, ed., *Towards Responsible Research and Innovation in the Information and Communication Technologies and Security Technologies Fields*, report from the European Commission Services (Luxembourg: Publications Office of the European Union, 2011).

50. Jack Stilgoe, Richard Owen, and Phil Macnaghten, "Developing a Framework for Responsible Innovation," *Research Policy* 42, no. 9 (2013): 1568–80.

51. John List and Fatemeh Momeni, "When Corporate Social Responsibility Backfires: Theory and Evidence from a Natural Field Experiment," National Bureau of Economic Research (NBER) Working Paper no. 24169, Cambridge, MA, December 2017.

52. Milton Friedman, "The Social Responsibility of Business Is to Increase Its Profits," *New York Times Magazine*, September 13, 1970.

53. Adam Thierer, "How Much Precaution Is Wise?" *Technology Liberation Front* (blog), November 1, 2019.

54. Bernd Carsten Stahl et al., "From Computer Ethics to Responsible Research and Innovation in ICT," *Information & Management* 51 (2014): 810–18.

55. Walter D. Valdivia and David H. Guston, "Responsible Innovation: A Primer for Policymakers," Center for Technology Innovation, Brookings Institution, May 2015.

56. Valdivia and Guston, "Responsible Innovation."

57. Florman, *The Existential Pleasures*, p. 22.

58. William McGeveran, "Friending the Privacy Regulators," *Arizona Law Review* 58 (2016): 979–80.

59. René von Schomberg, "The Precautionary Principle: Its Use within Hard and Soft Law," Symposium on the European Parliament's Role in Risk Governance, *European Journal of Risk Regulation* (2012): 154.

60. Schomberg, "The Precautionary Principle."

61. McGeveran, "Friending the Privacy Regulators": 987.

62. Philip J. Weiser, "Entrepreneurial Administration," *Boston University Law Review* 97 (2017): 2017.

63. Taylor Owen, *Disruptive Power: The Crisis of the State in the Digital Age* (Oxford: Oxford University Press, 2015), p. 191.

64. Calestous Juma, *Innovation and Its Enemies: Why People Resist New Technologies* (New York: Oxford University Press, 2016), p. 282.

Chapter 7

1. Gary E. Marchant and Braden Allenby, "New Tools for Governing Emerging Technologies," *Bulletin of the Atomic Scientists* 73 (2017): 108.

2. Kenneth W. Abbott, Gary E. Marchant, and Elizabeth A. Corley, "Soft Law Oversight Mechanisms for Nanotechnology," *Jurimetrics* 52, no. 3 (Spring 2012): 285.

3. Ryan Hagemann, "New Rules for New Frontiers: Regulating Emerging Technologies in an Era of Soft Law," *Washburn Law Journal* 57, no. 2 (Spring 2018): 249.

4. Wendell Wallach and Gary Marchant, "Toward the Agile and Comprehensive International Governance of AI and Robotics," *Proceedings of the IEEE* 107, no. 3 (March 2019): 506.

5. Julie E. Cohen, "The Regulatory State in the Information Age," *Theoretical Inquiries in Law* 17 (2016): 399.

6. Philip J. Weiser, "Entrepreneurial Administration," *Boston University Law Review* 97 (2017).

7. Cristie Ford, *Innovation and the State: Finance, Regulation, and Justice* (Cambridge: Cambridge University Press, 2017), p. 55.

8. William D. Eggers, Mike Turley, and Pankaj Kishnani, "The Regulator's New Toolkit: Technologies and Tactics for Tomorrow's Regulator," Deloitte Center for Government Insights, *Deloitte Insights* (October 18, 2018): 14.

9. Eggers, Turley, and Kishnani, "The Future of Regulation: Principles for Regulating Emerging Technologies," Deloitte Center for Government Insights, *Deloitte Insights* (June 19, 2018).

10. Ryan Hagemann, Jennifer Skees, and Adam D. Thierer, "Soft Law for Hard Problems: The Governance of Emerging Technologies in an Uncertain Future," *Colorado Technology Law Journal* 17, no. 1, (2016).

11. Hagemann, "New Rules for New Frontiers": 248.

12. White House, *The Framework for Global Electronic Commerce* (July 1997).

13. *The Framework for Global Electronic Commerce.*

14. "Management of Internet Names and Addresses, National Telecommunications and Information Administration, Department of Commerce," *Federal Register* 63, no. 111 (June 10, 1998): 31741.

15. Ira S. Rubinstein, "Regulating Privacy by Design," *Berkeley Technology Law Journal* 26 (2011): 1409.

16. Deborah G. Johnson, "Software Agents, Anticipatory Ethics, and Accountability," in *The Growing Gap Between Emerging Technologies and Legal-Ethical Oversight: The Pacing Problem,* ed. Gary E. Marchant, Braden R. Allenby, and Joseph R. Herkert (Dordrecht, Neth.: Springer, 2011), p. 64.

17. Adam Thierer, "Five Online Safety Task Forces Agree: Education, Empowerment & Self-Regulation Are the Answer," Progress & Freedom Foundation, *Progress on Point* 16, no. 13 (July 2009).

18. Federal Trade Commission, "Big Data: A Tool for Inclusion or Exclusion? Understanding the Issues," report, January 2016; Executive Office of the President, "Preparing for the Future of Artificial Intelligence," October 2016; Executive Office of the President, "Big Data: A Report on Algorithmic Systems, Opportunity, and Civil Rights," May 2016.

19. Federal Trade Commission, "Internet of Things: Privacy and Security in a Connected World," staff report, January 2015; Federal Trade Commission, "Careful Connections: Building Security in the Internet of Things," brochure, January 2015.

20. Federal Trade Commission, "Native Advertising: A Guide for Businesses," Tips & Advice, December 2015.

21. Department of Transportation, "Federal Automated Vehicle Policy: Accelerating the Next Revolution in Roadway Safety," guidance document, September 2016.

22. National Highway Traffic Safety Administration, "Cybersecurity Best Practices for Modern Vehicles," Report no. DOT HS 812 333, October 2016.

23. Food and Drug Administration, "Postmarket Management of Cybersecurity in Medical Devices," guidance document, December 28, 2016.

24. National Telecommunications and Information Administration, "Privacy Best Practice Recommendations for Commercial Facial Recognition Use," guidance document, June 15, 2016.

25. Food and Drug Administration, "Mobile Medical Applications: Guidance for Industry and Food and Drug Administration Staff," guidance document, February 9, 2015.

26. Food and Drug Administration, "Guidance for Industry: Internet/Social Media Platforms with Character Space Limitations—Presenting Risk and Benefit Information for Prescription Drugs and Medical Devices," draft guidance, June 2014.

27. Federal Trade Commission, "Mobile Privacy Disclosures: Building Trust Through Transparency," staff report, February 2013.

28. Federal Trade Commission, "Mobile Apps for Kids: Disclosures Still Not Making the Grade," survey report, December 2012.

29. Food and Drug Administration, "Technical Considerations for Additive Manufactured Medical Devices: Guidance for Industry and Food and Drug Administration Staff," guidance document, December 5, 2017.

30. Federal Aviation Administration, "Small Unmanned Aircraft Systems," Advisory Circular, AC no. 107-2, June 21, 2016; Federal Aviation Administration, "UAS Integration Pilot Program," Unmanned Aircraft Systems—Programs, Partnerships & Opportunities, November 7, 2018.

31. Ryan Hagemann, Jennifer Huddleston Skees, and Adam Thierer, "'Soft Law' Is Eating the World: Driverless Car Edition," Mercatus Center at George Mason University, *The Bridge*, October 11, 2018.

32. Adam Thierer et al., "How the Internet, the Sharing Economy, and Reputational Feedback Mechanisms Solve the 'Lemons Problem,'" working paper, Mercatus Center at George Mason University, Arlington, VA, May 2015.

33. Ian Sample, "Strongest Opponents of GM Foods Know the Least but Think They Know the Most," *The Guardian*, January 14, 2019.

34. Adam Thierer, "Five Online Safety Task Forces Agree."

35. Lynn Nielsen-Bohlman, Allison M. Panzer, and David A. Kindig, eds., Institute of Medicine of the National Academies, *Health Literacy: A Prescription to End Confusion* (Washington: National Academies Press, 2004). See also Scott C. Ratzan and Ruth M. Parker, "Introduction," in *National Library of Medicine Current Bibliographies in Medicine: Health Literacy*, ed. C. R. Selden et al., NLM Pub. no. CBM 2000-1 (Bethesda, MD: National Institutes of Health, U.S. Department of Health and Human Services, 2000).

36. Food and Drug Administration, *Strategic Plan for Risk Communication*, report, 2009.

37. Food and Drug Administration, *Communicating Risks and Benefits: An Evidence-Based User's Guide*, ed. Baruch Fischhoff, Noel T. Brewer, and Julie S. Downs, 2011.

38. See, for example, the FDA's Tools & Educational Materials webpages (http://www.fda.gov/food/fooddefense/toolseducationalmaterials/default .htm) and Educational Materials for Students & Teachers (https://www.fda .gov/food/resources-you-food/students-teachers) as well as the materials published by the FDA's Division of Industry and Consumer Education (https:// www.fda.gov/medical-devices/device-advice-comprehensive-regulatory -assistance).

39. Adam Thierer, "The Right to Try and the Future of the FDA in the Age of Personalized Medicine," working paper, Mercatus Center at George Mason University, Arlington, VA, July 2016, pp. 15–18.

40. Jordan Reimschisel and Adam Thierer, "A Model Roadmap for Genome Education," *Plain Text*, October 3, 2017.

41. Paul R. Marantz, "Rethinking Dietary Guidelines," *Critical Reviews in Food Science and Nutrition* 50 (December 2010): 17–18; Jill Carroll, "The Government's Food Pyramid Correlates to Obesity, Critics Say," *Wall Street Journal*, June 13, 2002.

42. Association for Computing Machinery, *Code of Ethics and Professional Conduct*, revised 2018.

43. *Code of Ethics and Professional Conduct.*

44. *Code of Ethics and Professional Conduct.*

45. The IEEE Global Initiative on Ethics of Autonomous and Intelligent Systems, *Ethically Aligned Design: A Vision for Prioritizing Human Well-Being with Autonomous and Intelligent Systems*, Version 2 (IEEE, 2017).

46. IEEE Global Initiative, p. 5.

47. Partnership for AI, "Building a Community of Practice: Reflections from Our 2nd All Partners Meeting," November 21, 2018.

48. OpenAI, "OpenAI Charter," accessed February 4, 2019, https:// blog.openai.com/openai-charter.

49. Hannah Devlin, "Do No Harm, Don't Discriminate: Official Guidance Issued on Robot Ethics," *The Guardian*, September 18, 2016.

50. British Standards Institution, "Robots and Robotic Devices: Guide to the Ethical Design and Application of Robots and Robotic Systems," no. 8611:2016, April 2016.

51. "About ISO," International Organization for Standardization (ISO), accessed January 13, 2017, http://www.iso.org/iso/home/about.htm.

52. "Developing Standards," ISO, accessed January 13, 2017, http:// www.iso.org/iso/home/standards_development.htm.

53. "13482:2014: Robots and Robotic Devices—Safety Requirements for Personal Care Robots," ISO, February 2014.

54. "Standards Catalogue: ISO/TC 299—Robotics," ISO, accessed January 13, 2017, http://www.iso.org/iso/home/store/catalogue_tc/catalogue _tc_browse.htm?commid=5915511.

55. Gregory N. Mandel, "Regulating Emerging Technologies," Temple University, Legal Studies Research Paper no. 2009-18 (2009): 9.

56. Gary E. Marchant and Wendell Wallach, "Governing the Governance of Emerging Technologies," in *Innovative Governance Models for Emerging Technologies*, ed. Gary E. Marchant, Kenneth W. Abbott, and Braden Allenby (Chelthenham, UK: Edward Elgar Publishing Limited, 2013), pp. 136–52.

57. Gary E. Marchant and Wendell Wallach, "Coordinating Technology Governance," *Issues in Science and Technology* (Summer 2015): 44–45.

58. Wendell Wallach and Gary Marchant, "Toward the Agile and Comprehensive International Governance of AI and Robotics," *Proceedings of the IEEE* 107, no. 3 (March 2019): 506.

59. Pam Belluck, "How to Stop Rogue Gene-Editing of Human Embryos?," *New York Times*, January 23, 2019.

60. Mary Douglas and Aaron Wildavsky, *Risk and Culture* (Berkeley: University of California Press, 1983), p. 194. See also William W. Lowrance, *Of Acceptable Risk: Science and the Determination of Safety* (Los Altos, CA: William Kaufmann, Inc., 1976).

61. Richard A. Posner, *Catastrophe: Risk and Response* (Oxford: Oxford University Press, 2004).

62. Adam Thierer, *Permissionless Innovation: The Continuing Case for Comprehensive Technological Freedom*, 2nd ed. (Arlington, VA: Mercatus Center at George Mason University, 2016), pp. 26–29; Aaron Wildavsky, *Searching for Safety* (New Brunswick, CT: Transaction Books, 1988), p. 38.

63. Adam Thierer, "Why Regulate Broadcasting: Toward a Consistent First Amendment Standard for the Information Age," Catholic University Law School *CommLaw Conspectus* 15 (Summer 2007): 431–82.

64. Adam Thierer, "The End of Censorship," *Technology Liberation Front* (blog), January 22, 2008.

65. Maureen K. Ohlhausen, "Painting the Privacy Landscape: Informational Injury in FTC Privacy and Data Security Cases," speech before the Federal Communications Bar Association, Washington, September 19, 2017.

66. Christopher Koopman et al., "Informational Injury in FTC Privacy and Data Security Cases," Mercatus Center, *Public Interest Comment*, October 27, 2017.

67. Adam Thierer, "The Pursuit of Privacy in a World Where Information Control Is Failing," *Harvard Journal of Law & Public Policy* 36 (2013).

68. Global Priorities Project, *Existential Risk: Diplomacy and Governance*, 2017, p. 6.

69. Nick Bostrom, "Existential Risks: Analyzing Human Extinction Scenarios and Related Hazard," *Journal of Evolution and Technology* 9, no. 1 (2002).

70. Nick Bostrom, "The Vulnerable World Hypothesis," working paper, V 3.15, 2018, p. 17.

71. Bostrom, "The Vulnerable World Hypothesis."

72. "Protocol for the Prohibition of the Use in War of Asphyxiating, Poisonous or Other Gases, and of Bacteriological Methods of Warfare," United Nations Office of Disarmament Affairs, June 17, 1925.

73. International Atomic Energy Agency, *Statute: As amended up to 28 December 1989*, p. 5, https://www.iaea.org/sites/default/files/statute.pdf.

74. International Atomic Energy Agency, *Statute*, p. 18.

75. International Atomic Energy Agency, *Statute*, pp. 18–19.

76. Kelsey Piper, "How Technological Progress Is Making It Likelier Than Ever That Humans Will Destroy Ourselves," *Vox*, November 19, 2018.

77. Robin Hanson, "Vulnerable World Hypothesis," *Overcoming Bias* (blog), November 16, 2018.

78. Campaign to Stop Killer Robots, "A Growing Global Coalition," accessed January 4, 2019, https://www.stopkillerrobots.org/about.

79. Future of Life Institute, "Lethal Autonomous Weapons Pledge," accessed January 4, 2019, https://futureoflife.org/lethal-autonomous-weapons-pledge.

80. Future of Life Institute, "Lethal Autonomous Weapons Pledge."

81. International Criminal Court, https://www.icc-cpi.int/about.

82. Pam Belluck, "How to Stop Rogue Gene-Editing of Human Embryos?," *New York Times*, January 23, 2019.

83. Adam Thierer, "Which Emerging Technologies Are 'Weapons of Mass Destruction'?," *Technology Liberation Front* (blog), August 26, 2016.

84. Eric S. Lander et al., "Adopt a Moratorium on Heritable Genome Editing," *Nature*, March 12, 2019.

85. Lander et al., "Adopt a Moratorium."

86. Amy Dockser Marcus, "Scientists Call for Moratorium to Block Gene-Edited Babies," *Wall Street Journal*, March 13, 2019.

87. Posner, *Catastrophe*, p. 81.

88. Thierer, *Permissionless Innovation*, pp. 33–38.

89. Henry Petroski, *To Engineer Is Human: The Role of Failure in Successful Design* (New York: Vintage, 1992), p. 105.

90. Peter Huber, "Exorcists vs. Gatekeepers in Risk Regulation," *Regulation* (November/December 1983): 28–29.

91. Posner, *Catastrophe*, p. 110.

92. Ban Ki-Moon, "Secretary-General's Remarks to Security Council Open Debate on the Non-Proliferation of Weapons of Mass Destruction," United Nations Secretary General, August 23, 2016.

93. Ban, "Secretary-General's Remarks to Security Council."

94. Ban, "Secretary-General's Remarks to Security Council."

95. Ban, "Secretary-General's Remarks to Security Council."

96. This section adapted from: Thierer, "Which Emerging Technologies Are 'Weapons of Mass Destruction'?"

97. Douglas Walton, *Fundamentals of Critical Argumentation* (Cambridge: Cambridge University Press, 2006), p. 285.

98. Walton, *Fundamentals of Critical Argumentation*, p. 287.

99. Adam Thierer, "Technopanics, Threat Inflation, and the Danger of an Information Technology Precautionary Principle," *Minnesota Journal of Law, Science & Technology* 14, no. 1 (2013).

100. Kenneth W. Abbott, "An International Framework Agreement on Scientific and Technological Innovation and Regulation," in *The Growing Gap Between Emerging Technologies and Legal-Ethical Oversight: The Pacing Problem*, ed. Gary E. Marchant, Braden R. Allenby, and Joseph R. Herkert (Dordrecht, Neth.: Springer, 2011), pp. 127–28.

101. Nick Bostrom, *Superintelligence: Paths, Dangers, Strategies* (Oxford: Oxford University Press, 2014), p. 259.

102. Brian Rappert, "Pacing Science and Technology with Codes of Conduct: Rethinking What Works," in *The Growing Gap Between Emerging Technologies and Legal-Ethical Oversight: The Pacing Problem*, ed. Gary E. Marchant, Braden R. Allenby, and Joseph R. Herkert, (Dordrecht, Neth.: Springer, 2011), pp. 109–26.

103. Milton L. Mueller, *Networks and States: The Global Politics of Internet Governance* (Cambridge, MA: MIT Press, 2010).

Chapter 8

1. Shoshana Zuboff, *The Age of Surveillance Capitalism: The Fight for a Human Future at the New Frontier of Power*, (New York: Public Affairs, 2019), p. 520.

2. Samuel C. Florman, *The Existential Pleasures of Engineering*, 2nd ed. (New York: St. Martin's Griffin, 1994), p. 72.

3. Robert D. Atkinson, Daniel Castro, and Alan McQuinn, "How Tech Populism Is Undermining Innovation," Information Technology and Innovation Foundation, April 1, 2015.

4. Adam D. Thierer, "Embracing a Culture of Permissionless Innovation," Cato Institute, Cato Online Forum, November 17, 2014, https://www .cato.org/publications/cato-online-forum/embracing-culture-permission less-innovation; Christopher Koopman, "Creating an Environment for Permissionless Innovation," Testimony before the U.S. Congress Joint Economic Committee, 115th Cong., 2nd sess., May 22, 2018.

5. Stephen Ezell and Philipp Marxgut, "Comparing American and European Innovation Cultures," in *Shaping the Future: Economic, Social, and Political Dimensions of Innovation* (Austrian Council for Research and Technology Development, August 2015); Virgil Storr, *Understanding the Culture of Markets* (New York: Routledge Foundations of the Market Economy, 2013), p. 1.

6. Maike Didero et al., "Differences in Innovation Culture Across Europe: A Discussion Paper," Transform Consortium, Bonn, February 2008, p. 3.

7. Joel Mokyr, *Lever of Riches: Technological Creativity and Economic Progress* (New York: Oxford University Press, 1990), pp. 16, 182.

8. Robert D. Atkinson, "Understanding the U.S. National Innovation System," Information Technology and Innovation Foundation, June 2014, p. 8.

9. Benjamin M. Friedman, "The Moral Consequences of Economic Growth," *Society* (January/February 2006): 16.

10. Alan Greenspan and Adrian Wooldridge, "How to Fix the Great American Growth Machine," *Wall Street Journal*, October 12, 2018. See also Ezell and Marxgut, "Comparing American and European Innovation Cultures," p. 193.

11. Deirdre N. McCloskey, *The Bourgeois Virtues: Ethics for an Age of Commerce* (Chicago: University of Chicago Press, 2006); Deirdre N. McCloskey, *Bourgeois Dignity: Why Economics Can't Explain the Modern World* (Chicago: University of Chicago Press, 2010).

12. Adam Thierer, "How Attitudes about Risk & Failure Affect Innovation on Either Side of the Atlantic," *Medium*, June 19, 2015.

13. Stace Lindsay, "Culture, Mental Models, and National Prosperity," in *Culture Matters: How Values Shape Human Progress*, ed. Lawrence Harrison and Samuel Huntington (New York: Perseus Books Group, 2000), p. 282.

14. David Landes, *The Wealth and Poverty of Nations: Why Some Are So Rich and Some Are So Poor* (New York: W. W. Norton, 1998), p. 516.

15. Paul Herbig and Steve Dunphy, "Culture and Innovation," *Cross Cultural Management: An International Journal* 5, no. 4 (1998): 14.

16. Mancur Olson, "Big Bills Left on the Sidewalk: Why Some Nations Are Rich, and Others Poor," *Journal of Economic Perspectives* 10, no. 2 (Spring 1996): 22, 19.

17. Elhanan Helpman, *The Mystery of Economic Growth* (Cambridge, MA: Belknap Press of Harvard University Press, 2004), p. 115.

18. Philippe Aghion, Ufuk Akcigit, and Peter Howitt, "Lessons from Schumpeterian Growth Theory," *American Economic Review* 105, no. 5 (May 2015): 5.

19. Helpman, *The Mystery of Economic Growth*, p. 140.

20. Patrick McLaughlin and Laura Stanley, "Regulation and Income Inequality: The Regressive Effects of Entry Regulations," working paper, Mercatus Center at George Mason University, Arlington, VA, January 2016.

21. Michael Porter, "Attitudes, Values, Beliefs, and the Microeconomics of Prosperity," in *Culture Matters: How Values Shape Human Progress*, ed. Lawrence Harrison and Samuel Huntington (New York: Perseus Books Group, 2000), p. 17.

22. White House, *The Framework for Global Electronic Commerce* (July 1997).

23. Neil Chilson, "Preserving Permissionless Innovation," *Morning Consult*, February 15, 2019.

24. Adam Thierer, "Bipartisan Internet of Things Resolution Introduced in Senate," *Technology Liberation Front* (blog) March 4, 2015.

25. "Expressing the Sense of the Senate about a Strategy for the Internet of Things to Promote Economic Growth and Consumer Empowerment," S. Res. 110, 114th Cong. (March 24, 2015).

26. "Expressing the Sense of the Senate."

27. Jerry Ellig, "Dynamic Competition, Online Platforms, and Regulatory Policy," Mercatus Center at George Mason University, *Public Interest Comment*, December 9, 2015, https://www.mercatus.org/publication/dynamic-competition-online-platforms-and-regulatory-policy.

28. Matthew D. Mitchell and Tad DeHaven, "Populism, Markets, and the Elite," Mercatus Center at George Mason University, *The Bridge*, February 1, 2019.

29. Dustin Chambers and Jonathan Munemo, "The Impact of Regulations and Institutional Quality on Entrepreneurship," Mercatus Center, *Mercatus Research*, (2017), p. 26.

30. Matthew D. Mitchell, "Occupational Licensing and the Poor and Disadvantaged," Mercatus Center policy brief, September 28, 2017.

31. Morris M. Kleiner and Evgeny S. Vorotnikov, *At What Cost? State and National Estimates of the Economic Costs of Occupational Licensing* (Arlington, VA: Institute for Justice, 2018), p. 5.

32. *North Carolina State Board of Dental Examiners v. FTC*, 135 S. Ct. 1101 (2015).

33. Federal Trade Commission, "Selected Advocacy Relating to Occupational Licensing," accessed February 8, 2019, https://www.ftc.gov/policy/advocacy/economic-liberty/selected-advocacy-relating-occupational-licensing. See also Matthew Mitchell and Ryan Nunn, "Rule Reversal: How the Feds Can Challenge State Regulation," *Wall Street Journal*, July 6, 2017.

34. Federal Trade Commission, "FTC Launches New Website Dedicated to Economic Liberty," press release, March 16, 2017.

35. Adam Thierer and Trace Mitchell, "A Non-partisan Way to Help Workers and Consumers," Mercatus Center at George Mason University, *The Bridge*, September 25, 2018.

36. Nick Sibilla, "New Kentucky Law Will Let Anyone Start a Home-Baking Business," *Forbes*, April 3, 2018.

37. John Chisholm, *Unleash Your Inner Company: Use Passion and Perseverance to Build Your Ideal Business* (Austin, TX: Greenleaf, 2015), p. 308.

38. Morris M. Kleiner, "Reforming Occupational Licensing Policies," Brookings Institution, The Hamilton Project, Discussion Paper 2015-01, January 2015, p. 20.

39. Adam Thierer, "Converting Permissionless Innovation into Public Policy: 3 Reforms," *Medium*, November 29, 2017.

40. Patrick McLaughlin and Michael Wilt, "Regulatory Accumulation: The Problem and Solutions," Mercatus Center *Policy Spotlight,* September 2017.

41. Laura Howard and James Broughel, "Trump's Regulatory Reforms Need a Phase Two," *The Hill*, September 24, 2018.

42. Philip K. Howard, *The Death of Common Sense: How Law Is Suffocating America* (New York: Random House, 1994, 2011), p. 203.

43. Lyn M. Gaudet and Gary E. Marchant, "Administrative Law Tools for More Adaptive and Responsive Regulation," in *The Growing Gap Between Emerging Technologies and Legal-Ethical Oversight: The Pacing Problem*, ed. Gary E. Marchant, Braden R. Allenby, and Joseph R. Herkert (Dordrecht, Neth.: Springer, 2011), p. 178.

44. Sofia Ranchordás, "Does Sharing Mean Caring? Regulating Innovation in the Sharing Economy," *Minnesota Journal of Law, Science & Technology* 16 (2015): 39.

45. Robert J. Thornton and Edward J. Timmons, "The De-Licensing of Occupations in the United States," U.S. Bureau of Labor Statistics, *Monthly Labor Review,* May 2015, p. 13.

46. Brian Baugus and Feler Bose, "Sunset Legislation in the States: Balancing the Legislature and the Executive," Mercatus Center at George Mason University, *Mercatus Research*, August 2015.

47. James Broughel, "A Reform That Offers Hope for Centrists," *Washington Post*, March 14, 2018.

48. Brian Knight, "Federalism and Federalization on the Fintech Frontier," *Vanderbilt Journal of Entertainment and Technology Law* 20, no. 1 (2017): 132.

49. Anupam Chander, *The Electronic Silk Road: How the Web Binds the World Together in Commerce* (New Haven, CT: Yale University Press, 2013), p. 154.

50. World Trade Organization, "Understanding the WTO," Geneva, Switzerland, 2015, p. 11.

51. Christopher Koopman, Matthew Mitchell, and Adam Thierer, "The Sharing Economy and Consumer Protection Regulation: The Case for Policy Change," *Journal of Business, Entrepreneurship & Law* 8, no. 2 (2015): 529–45.

52. Kleiner, "Reforming Occupational Licensing Policies," p. 20.

53. Charles Murray, *By the People: Rebuilding Liberty without Permission* (New York: Crown Forum, 2016), p. 17.

54. Adam Thierer and Eli Dourado, "The Innovator's Defense Fund: What It Is and Why It's Needed," *Medium*, February 25, 2016.

55. Alana Abramson, "A New VC Firm Tied to the Koch Brothers Plans to Take on Government Regulations," *Fortune*, January 31, 2018.

56. Robin I. Mordfin, "Representing the Bleeding Edge: UChicago Law's New Innovation Clinic," University of Chicago Law School, March 31, 2016.

57. Creative Destruction Lab, "Increase Probablity of Success," https://www.creativedestructionlab.com/program/.

58. Tusk Ventures, "What We Do," accessed July 11, 2018, https://tuskventures.com/about.

59. Tusk Ventures, "What We Do."

60. Institute for Justice, *Entrepreneur's Survival Guide*, advisory report, September 2014.

61. Anthony Luppino, "How Law Schools Can and Should Be Involved in Building Ecosystems that Foster Innovation, Entrepreneurship, and Growth," Ewing Marion Kauffman Foundation.

62. Julian Simon, *The Ultimate Resource 2* (Princeton, NJ: Princeton University Press, 1996), p. 589.

63. Herman Kahn and Anthony Wiener, *The Year 2000: A Framework for Speculation on the Next Thirty-Three Years* (New York: Macmillan, 1967), pp. 264–65.

64. Jennifer Huddleston Skees, "Should You Be Able to Fix Your Own iPhone?," *Plain Text*, June 18, 2018.

65. Cristie Ford, *Innovation and the State: Finance, Regulation, and Justice* (Cambridge: Cambridge University Press, 2017), p. 49.

66. FAA, "Fact Sheet—Commercial Aviation Safety Team," April 12, 2016.

67. Richard Williams, "Regulators Must Become as Creative as Innovators," *The Hill*, December 13, 2018.

68. Ryan Hagemann, "New Rules for New Frontiers: Regulating Emerging Technologies in an Era of Soft Law," *Washburn Law Journal* 57, no. 2 (Spring 2018): 255.

69. Hagemann, "New Rules for New Frontiers," p. 244.

Chapter 9

1. Joe Carlen, *A Brief History of Entrepreneurship: The Pioneers, Profiteers, and Racketeers Who Shaped Our World* (New York: Columbia University Press, 2016), p. 215.

2. Eli Dourado, "How Technological Innovation Can Massively Reduce the Cost of Living," *Plain Text*, January 29, 2016.

3. Niklas Elert and Magnus Henrekson, "Evasive Entrepreneurialism," *Small Business Economics* 47 (2016): 95.

4. Adam Thierer, *Permissionless Innovation: The Continuing Case for Comprehensive Technological Freedom*, 2nd ed. (Arlington, VA: Mercatus Center at George Mason University, 2016), p. 2.

5. Matt Ridley, *The Rational Optimist: How Prosperity Evolves* (New York: Harper Collins, 2010).

6. Virginia Postrel, *The Future and Its Enemies* (New York: The Free Press, 1998), p. 212.

7. F. A. Hayek, *The Constitution of Liberty* (London: Routledge, 1960, 1990), pp. 85, 87.

8. Steven Horwitz and Jack Knych, "The Importance of Failure," *Freeman* 61, no. 9 (November 2011).

9. Samuel C. Florman, *The Existential Pleasures of Engineering*, 2nd ed. (New York: St. Martin's Griffin, 1994), p. 84.

10. Jonathan Klick and Gregory Mitchell, "Government Regulation of Irrationality: Moral and Cognitive Hazards" *Minnesota Law Review* 90 (2006): 1626.

11. Adam Thierer, "Failing Better: What We Learn by Confronting Risk and Uncertainty," in *Nudge Theory in Action: Behavioral Design in Policy and Markets*, ed. Sherzod Abdukadirov (New York: Palgrave Macmillan, 2016): 65–94.

12. Mary Douglas and Aaron Wildavsky, *Risk and Culture* (Berkeley: University of California Press, 1983), p. 195.

13. Ryan Hagemann, "'Let's Have a Conversation' about Technology: Policy Punting or Crypto-Luddite Deflection?," *Niskanen Center* (blog), September 20, 2017.

14. Ian Barbour, *Ethics in an Age of Technology* (New York: Harper One, 1993), p. xvii.

15. Brett Frischmann and Evan Selinger, *Re-Engineering Humanity* (Cambridge: Cambridge University Press, 2018), p. 241.

16. Adam Thierer, "What Does It Mean to 'Have a Conversation' about a New Technology?," *Technology Liberation Front* (blog), May 23, 2013.

17. Karl Popper, *The Poverty of Historicism* (London: Routledge Classics, 1957), p. 77.

18. F. A. Hayek, "The Use of Knowledge in Society," *The American Economic Review* 35, no. 4 (September 1945): 519.

19. Alan McQuinn, "From Kodak To Google, How Privacy Panics Distort Policy," *TechCrunch*, October 2, 2015, https://techcrunch.com/2015/10/01 /from-kodak-to-google-how-privacy-panics-distort-policy.

20. Samuel D. Warren and Louis D. Brandeis, "The Right to Privacy," *Harvard Law Review* 4 (1890): 193. (In which the authors lamented that "instantaneous photographs and newspaper enterprise have invaded the sacred precincts of private and domestic life" and claimed that "numerous mechanical devices threaten to make good the prediction that 'what is whispered in the closet shall be proclaimed from the house-tops.'")

21. Thierer, *Permissionless Innovation*, p. 2. See also Adam Thierer, "Muddling Through: How We Learn to Cope with Technological Change," *Medium*, June 30, 2014; Florman, *The Existential Pleasures*, p. 84.

22. James C. Scott, *Seeing Like a State: How Certain Schemes to Improve the Human Condition Have Failed* (New Haven, CT: Yale University Press, 1998), p. 313.

23. Kathleen Waugh and Gary E. Marchant, "Collaborative Voluntary Programs: Lessons from Environmental Law," in *The Growing Gap Between Emerging Technologies and Legal-Ethical Oversight: The Pacing Problem*, ed. Gary E. Marchant, Braden R. Allenby, and Joseph R. Herkert (Dordrecht, Neth.: Springer, 2011), p. 193.

24. Chris Freeman and Luc Soete, *Economics of Industrial Innovation*, 3rd ed. (London: Routledge, 1997), p. 194.

25. Peter L. Bernstein, *Against the Gods: The Remarkable Story of Risk* (New York: John Wiley & Sons, 1996), p. 334.

26. Elinor Ostrom, "Beyond Markets and States: Polycentric Governance of Complex Economic Systems." *American Economic Review* 100, no. 3 (2009): 641–72.

27. Paul D. Aligica and Vlad Tarko, "Polycentricity: From Polanyi to Ostrom, and Beyond," *Governance: An International Journal of Policy, Administration, and Institutions* 25, no. 2 (April 2012): 237–62.

28. Joshua S. Goldstein and Staffan A. Qvist, "Only Nuclear Energy Can Save the Planet," *Wall Street Journal*, January 11, 2019.

29. Max Roser, "Most of Us Are Wrong about How the World Has Changed (Especially Those Who Are Pessimistic about the Future," *Our World in Data* (blog), July 27, 2018; Bobby Duffy, *The Perils of Perception: Why We're Wrong about Nearly Everything* (London: Atlantic Books, 2018).

30. Calestous Juma, *Innovation and Its Enemies: Why People Resist New Technologies* (New York: Oxford University Press, 2016), p. 289.

31. Arthur M. Diamond Jr., *Openness to Creative Destruction: Sustaining Innovative Dynamism* (Oxford: Oxford University Press, 2019).

32. Samuel C. Florman, *Blaming Technology: The Irrational Search for Scapegoats* (New York: St. Martin's Press, 1981), p. 193.

Postscript: Evasiveness during a Pandemic

1. Peggy Noonan, "'Don't Panic' Is Rotten Advice," opinion, *Wall Street Journal*, March 12, 2020.

2. Adam Thierer, "How the US Botched Coronavirus Testing," American Institute for Economic Research, March 12, 2020.

3. Neel V. Patel, "Why the CDC Botched Its Coronavirus Testing," *MIT Technology Review*, March 5, 2020.

4. Sheri Fink and Mike Baker, "'It's Just Everywhere Already': How Delays in Testing Set Back the U.S. Coronavirus Response," *New York Times*, March 10, 2020.

5. Fink and Baker, "'It's Just Everywhere Already.'"

6. Fink and Baker.

7. Suzy Khimm, Laura Strickler, and Brenda Breslauer, "Many Private Labs Want to Do Coronavirus Tests. But They're Still Facing Obstacles and Delays," *NBC News*, March 11, 2020.

8. Fink and Baker, "'It's Just Everywhere Already.'"

9. "Johns Hopkins Develops Its Own Coronavirus Test," *CBS News*, March 13, 2020.

10. Trevor Smale, "Open Source Low Resource Ambu-Bag Ventilator," GitLab, Project ID: 17454594, accessed March 16, 2020.

11. Davide Sher, "Italian Hospital Saves Covid-19 Patients Lives by 3D Printing Valves for Reanimation Devices," *3D Printing Media Network*, March 14, 2020.

12. Albert Han, "The Polytechnic University Lab 3D Printing Face Shields for Coronavirus-Battling Hong Kong Hospital Workers," *South China Morning Post*, February 25, 2020.

13. Jacquie Lee, "Hand Sanitizer Shortages Push FDA to Let Pharmacists Make It," *Bloomberg*, March 14, 2020.

14. Dan Kois, "America Is a Sham," *Slate*, March 14, 2020.

15. Colin A. Young and Matt Murphy, "Coronavirus Testing Limitations a Growing Concern for Mass. Gov. Baker," *New England Public Radio*, March 12, 2020.

Note: Page numbers followed by "f" or "t" indicate figures and tables, respectively. Page numbers with "n" or "nn" indicate notes.

accountability of government,
159–81
and continuation of tech-state
power game, 179–81
dissent's role in enforcing,
117–19, 123
and failure of reforms to
constrain regulation, 160–62
innovation's role in, 13, 158,
168–75, 266
technology as constrainer vs.
expander of government,
162–66
and technology as eroder of
nation-state, 166–68
voice and exit power of
innovators, 175–79
ACM. *See* Association of Computing Machinery (ACM)
adaptation in face of adversity,
importance of innovation to,
184–86, 271–73

adaptive regulation, 208. *See also*
soft law
Adler, Jonathan, 137
agency discretion or forbearance.
See rule departure by
government
AI. *See* artificial intelligence
Airbnb, 62–63, 64, 178
Allenby, Braden, 74
alternative transportation
services, 59–64, 70, 92–94,
178–79
Aman, Alfred, 67, 180
Amazon, drone innovation and
regulatory restrictions, 67–68
anti-technology movement,
20–21, 274–75
antitrust legislation, as support for
new-entry innovation, 249
Arab Spring uprising, and
disruption of hegemons, 169
Arendt, Hannah, 115, 116, 117, 150

Arthur, W. Brian, 34
artificial intelligence (AI), 22, 38, 50, 71, 74, 211, 217, 220–21, 237
Association of Computing Machinery (ACM), 219
automobile industry
car dealerships, and crony capitalism, 144
driverless car challenge to, 70, 94–98, 127, 128, 139, 309n12
problem of obsolete laws and regulations for, 140–41
protectionism by, 144
autonomous vehicles. *See* driverless cars
autonomous weapons, seeking international control over, 233
aviation and space innovation, regulatory restrictions on, 106–7. *See also* drones

Ban Ki-Moon, 236–37, 238
Barlow, John Perry, 165
Barry, Jordan M., 59
Baumol, William, 40–41, 291n6
Being Digital (Negroponte), 165
Benkler, Yochai, 168–69
Bernstein, Peter L., 272
biohacking, 79–87
Bitcoin, 50, 98–99, 100, 172, 305n252
Blaming Technology: The Irrational Search for Scapegoats (Florman), 275

blockchain technology, 98–103, 156, 172, 305n252
Boettke, Peter, 55
"born free" vs. "born captive" entrepreneurial sectors, 104–7, 138–39
Bostrom, Nick, 228–32
Boudreaux, Donald, 34, 46
bounded rationality model, 216
Bredderman, Will, 60–61
Brief History of Entrepreneurship (Carlen), 41, 265
British Standards Institute, 221
Broughel, James, 35, 254
Brunton, Finn, 112–13
build-and-freeze rules from risk-averse regulators, 135–42
bureaucracy
federalism as check on, 161
legislative and executive inability to control, 139, 151–53, 162
Burfield, Evan, 64–65
Buttarelli, Giovanni, 26
By the People: Rebuilding Liberty without Permission (Murray), 113, 257

The Captured Economy: How the Powerful Enrich Themselves, Slow Down Growth, and Increase Inequality (Lindsey and Teles), 143
Carbonara, Emanuela, 116–17

car dealerships, and crony
    capitalism, 144
Carlen, Joe, 41, 265
Castro, Daniel, 49
CFTC. *See* Commodity Futures
    Trading Commission
    (CFTC)
Chambers, Dustin, 52, 246–47
Chander, Anupam, 255
Chao, Elaine, 127
Chaturvedi, Swati, 46
Chisholm, John, 250
Churi, Salen, 258
Citymapper, 93–94
civil disobedience. *See also* techno-
    logical civil disobedience
    defined, 114
    libertarian motivation for,
        113
    profit seeking and, 119–22
    reasons for, 110–12
    risks for those engaged in,
        121–22
    social justice motivation for,
        112–13
Clinton, Bill, 209, 245–46
collaborative regulation, 208–10
Collins, Sonya, 81
combinatorial innovation, 8
Comma.ai, 97
commercial vs. noncommercial
    activities, and civil
    disobedience, 119–22
commercial vs. social
    entrepreneurialism, 47–48

Commodity Futures Trading
    Commission (CFTC), 101
common law mechanisms for
    technological governance, 12
communications and media
    policy, 72–73, 136, 146, 251
competency trap, 9
competition
    crony capitalism as subverter
        of healthy, 143–50
    as gateway to innovative
        freedom, 248–50
    as motive for established firms
        to support inflexible
        regulations, 142
competitive federalism, 16, 70–71
compliance paradox, 10, 16, 262
Congress
    abrogation of responsibilities
        to federal bureaucracy, 139,
        151–52, 162
    moves toward permissionless
        innovation, 68, 73, 246
    resistance to innovation, 127, 150
    vulnerability to special interest
        lobbying, 149
Consumer Product Safety
    Commission, 199
consumers, freedom to innovate
    as benefit for, 260
co-regulation, 207
corporate social responsibility
    (CSR), 197
Cortez, Michelle, 82
cottage food entrepreneurs, 88

Cowen, Tyler, 102, 143, 163, 167

Coyne, Christopher, 55

creative destruction, importance to innovation, 40, 75, 143, 144, 244

Crews, Wayne, 132–33

criminal offenses, proliferation of, 134–35

CRISPR, 86–87

crony capitalism, 62, 143–50

cryptocurrencies, 98–100, 172

CSR. *See* corporate social responsibility (CSR)

culinary entrepreneurialism, 87–90

A Culture of Improvement: Technology and the Western Millennium (Friedel), 44

Cuomo, Andrew, 61

cybersecurity, governmental focus on, 26

A Dangerous Master: How to Keep Technology from Slipping beyond Our Control (Wallach), 153

darknet transactions, 102, 172

Davidson, James Dale, 67

The Death of Common Sense (Howard), 132

de Blasio, Bill, 60, 61

decentralized marketplaces, 98–103

decentralized medicine, 84–87

deep technologies, defined, 46

degrowth movement, 21–22

Deleting Online Predators Act, 25

delivery robots, 70–71

demand for and fear of novelty, 28–33

democracy
importance of civil disobedience to, 115–16
and innovation as promoter of freedom, 42, 43–44, 113, 266
technology as disruptive tool to return power to people, 174–75

demosclerosis, 16, 150–53

dental aligners, 82–83

Department of Transportation (DOT), 127

determinism, technological, 186–93

digital citizens, improving education of, 216

Digital Health Innovation Action Plan of FDA, 76

Disruptive Power: The Crisis of the State in the Digital Age (Owen), 166

dissent
as enforcement of accountability, 117–19, 123
innovation as agent of, 7–8
value of American tradition, 113–14

DIY (do it yourself) medical technologies, 79–87

DOT. *See* Department of Transportation (DOT)

Dourado, Eli, 177, 257–58

Downes, Larry, 153

driverless cars, 70, 94–98, 127, 128, 139, 309n12

drones
  advantages over larger aircraft in freedom to operate, 107
  as challenge to archaic laws, 131
  commercial vs. amateur use, 126–27
  innovation arbitrage caused by regulation, 67–69
  pressure to reform regulation, 73, 90–92
  restrictions of born-in-captivity status, 139

dropgangs, 102

Dudley, Amos, 82–83

economic growth
  entrepreneurialism's role in, 41–42
  importance of competition and new entry for, 246–47
  innovation as driver of, 29, 34–35, 285n103
  as moral imperative, 41–44
  negative impact of crony capitalism on, 143

questioning of benefits from, 22
  regulatory accumulation's effect on, 133–34

The Economics of Regulation (Alfred Kahn), 148–49

Edelman, Benjamin, 103

Ehrenkranz, Melanie, 126

electric scooter rental, 93

Elert, Niklas, 123

Elwell, Dan, 140

Enabling the Future (e-NABLE), 78

energy industry, regulatory restrictions on, 106

entrepreneurial administration, 127–28, 262–63

entrepreneurialism
  born free vs. born captive sectors, 104–7, 138–39
  commercial vs. social entrepreneurialism, 47–48
  contribution to human betterment, 3–13, 34, 41–43
  economic growth role of, 41–42
  legal defense strategies to support entrepreneurs, 257–29
  political, 55, 143–50
  risk-taking culture of, 40–41
  social, 47–48

Ethereum network, 101

evasive entrepreneurialism
  defense fund for entrepreneurs, 257–59
  defined, 1–2, 16, 54–55

ex ante vs. ex post question for, 58–59, 122–24, 214–15, 227–28

and global economy's freedom of location, 66–67

motivations of, 55–56

overview of benefits, 3–13

regulatory entrepreneurs, 17, 54–57, 59–65

as response to demosclerosis and kludgeocracy, 151–53

rise of, 129–58

spontaneous deregulation, 103–4, 106

use of voice and exit to push change, 176–78

ex ante vs. ex post regulatory remedies, 58–59, 122–24, 214–15, 227–28

existential risks of technology, dealing with, 228–36

Exit, Voice, and Loyalty (Hirschman), 175–76

export controls on new technology, 74–75

FAA. *See* Federal Aviation Administration (FAA)

FAA Reauthorization Bill of 2018, 68

Facebook, and digital currency, 100

fallacy of inconsistency, 24

false equivalence in anti-technology arguments, 24, 26–27, 237–38

FCC. *See* Federal Communications Commission (FCC)

FDA. *See* Food and Drug Administration (FDA)

fear appeals from tech critics, 23–28, 237–38

fear of innovation until new is demanded, 28–33

Federal Aviation Administration (FAA)

and drone technology challenge, 67–69, 90–91, 126–27, 131, 139, 153–54

resistance to new technologies, 136

Federal Communications Commission (FCC), 72–73, 136, 146, 251

federalism, as check on unbridled bureaucracy, 161

Federal Trade Commission (FTC), 62, 226, 249

Fellow, Anthony R., 92

50 Things That Made the Modern Economy (Harford), 37

financial technology (fintech), as born regulatory captive, 106

firearms, 3D-printed, 78–79

flexible governance, advantages of, 272–74

flexible regulation. *See* soft law

floating transport, 93–94

Florman, Samuel C., 22, 164, 188, 189, 193, 274–75

Foer, Franklin, 22

Food and Drug Administration (FDA)
common-sense approach to precaution, 199
genetic testing, 65, 69–70
health and fitness smartphone apps, 128
medical device approval, 76–77, 81–82, 120–21
risk-averse culture at, 137–38
risk education role, 217
rule departure by, 127
struggle to keep up with innovation, 154

food and plant entrepreneurialism, 87–90

FoodTech Revolution, 87–90

food trucks, 87–88

Ford, Cristie, 20, 136, 155, 173, 208

forecasting, risks of technological, 272

for-profit activities, and technological civil disobedience, 56

Four Thieves Vinegar, 84–85

Framework for Global Electronic Commerce, 209, 245–46

Franklin, Benjamin, 111

freedom, innovation as promoter of, 42, 43–44, 113, 266

freedom of association and spirit of resistance, 115–17

free innovation, 16, 47–48, 56, 97–98, 260, 261–62

Friedel, Robert, 44

Friedman, Benjamin, 5, 244

Friedman, Milton, 66, 197

FTC. *See* Federal Trade Commission (FTC)

Future Crimes (Goodman), 22

future shock, 23

gambling outposts, distributed, 101

GCCs. *See* governance coordinating committees (GCCs)

genetically modified organisms (GMOs), 24–25, 29, 216

genetic innovations, 65, 69, 72, 86–87, 154, 155f, 233, 234

Geradin, Damien, 103

Gilder, George, 165

globalization
competition for innovation, 71–75
and disruptive effect of change on governments, 180
and erosion of nation-state, 167–68
innovation arbitrage effects, 66–68

GMOs. *See* genetically modified organisms (GMOs)

Goodman, Marc, 22

Google Glass tech failure, 189–90

governance coordinating committees (GCCs), 223, 224, 239

government. *See also* accountability of government; technological governance
constructive response to technological civil disobedience, 9–12
cybersecurity focus of, 26
globalization's effect on, 180
health literacy role of, 217–18
need for rules to govern markets and innovation, 58–59
overview, 6–8
rule departure, 103, 124–28
technology's empowerment of, 163–66, 170

Government's End: Why Washington Stopped Working (Rauch), 151

grassroots innovation. *See* free innovation

Greenspan, Alan, 244

Grindfest, 85–86

Gulfo, Joseph V., 138

Gurri, Martin, 168–69

Guston, David H., 200

Hagemann, Ryan, 156, 209, 263

Hammond, Samuel, 180

hard vs. soft law, 206, 213, 224–35

Harford, Tim, 37

harm from innovation, challenge of defining, 224–28

Haskell, Thomas L., 194–95

Hayek, F. A., 268

health care. *See also* medical technologies
cost of delays in adopting new developments, 137–38
medical tourism as innovation arbitrage, 71–72
pacing problem for, 154, 156
risk education for, 217–18

health literacy, government role in, 217–18

hegemons, 167, 168

Heinla, Ahti, 71

Helpman, Elhanan, 245

Henrekson, Magnus, 123

Hern, Alex, 93–94

Hirschman, Albert O., 13, 175–76, 181

Holmes, Elizabeth, 123

Hopper, Grace M., 57

Hopper's Law, 57

Hotz, George, 97

hours of labor needed to purchase household items, 36–38, 38t

Howard, Philip K., 132, 134, 149, 253

Huddleston, Jennifer, 209

Huerta, Michael, 153, 262

Hughes, Thomas P., 165

human betterment and
flourishing
and adaptation in face of
adversity, 184–86, 271–73
entrepreneurialism's contribution to, 3–13, 34, 41–43
importance of freedom to
humanity, 266–67
innovation's contribution to,
31–40, 33f, 43–44, 45–46, 45f
risk taking as contributor to,
3–13, 51
technology as expander of
horizons, 193–96
technology as historically
consistent agent of, 192
humanism
moral foundation of, 194, 195
relationship to technology,
190–91
humanitarian sensibility,
technology's effect on,
194–96
*Hybrid: The History and Science
of Plant Breeding*
(Kingsbury), 29

ICANN. *See* Internet Corporation of Assigned Names and
Numbers (ICANN)
IEEE. *See* Institute of Electrical
and Electronics Engineers
(IEEE)
infant mortality, reduction in, 36

information and culture,
improvements from
innovation, 36
innovation. *See also* evasive
entrepreneurialism
contribution to human
betterment, 31–40, 33f,
43–44, 45–46, 45f
cost-benefit analysis, 20–21
creative destruction's importance to, 40, 75, 143, 144, 244
culture of, 17, 243–46
as driver of economic growth,
29, 34–35, 285n103
entrepreneurialism and risk as
necessary to, 40–44
historical perspective on
disruptive technologies,
29–32
importance to human adaptation, 184–86, 271–73
iteration vs., 44–47
moral dimension of. *See* moral
dimension of innovation
permissionless. *See* permissionless innovation
as promoter of freedom, 42,
43–44, 113, 266
as public policy priority, 48–52
role in accountability of
government, 13, 158,
168–75, 266
as voice and exit, 13, 114–15,
175–79

Innovation and Its Enemies: Why People Resist New Technologies (Juma), 20, 57, 274

Innovation and the State (Ford), 173, 208

innovation arbitrage, 65–75
challenge for financial technology, 106
defined, 16
driverless cars, 97, 98
and existential risks of innovation, 234–35
globalization's effect on, 66–68
increasing opportunities for, 180
medical tourism as, 71–72
strike back from government alliances against, 180–81
and U.S. federalism, 161

Innovation Breakdown: How the FDA and Wall Street Cripple Medical Advances (Gulfo), 138

innovative dynamism, political favoritism as damper on, 143

Innovator's Defense Fund, 257–59

Innovator's Presumption, 250–52

Institute of Electrical and Electronics Engineers (IEEE), 219, 220

Internal Revenue Service (IRS), 100

International Congress for the Governance of AI, proposal for, 223

international environment. *See also* globalization
existential risks of technology, 228–36
robotics, ethical codes and international agreements, 221, 222, 233
technological governance in, 223–24, 228–35, 238–39

internet
as born free to innovate, 138–39
collaborative approach to policy on, 209–10
crony capitalism's ineffectiveness in controlling, 146–47
failure to create utopia of individual freedom, 164–66
governmental surveillance use of, 164, 165–66, 170

Internet Corporation of Assigned Names and Numbers (ICANN), 239

Internet of Things, 72, 246

IRS. *See* Internal Revenue Service (IRS)

ISO, 221–22

iteration vs. disruptive innovation, 44–47

Jefferson, Thomas, 115–16, 163–64

Jorgenson, Jillian, 60–61

journalistic uses of drones, 91

judicial branch, failure to meaningfully regulate regulators, 162
Juma, Calestous, 20, 29, 57, 274
jurisdictional competition. *See* innovation arbitrage

Kahn, Alfred, 148–49
Kahn, Herman, 261
Kalanick, Travis, 61, 63
King, Martin Luther Jr., 111
Kingsbury, Noel, 29, 32
Kirzner, Israel, 56
Kleiner, Morris, 147
kludgeocracy, 150–53
Knight, Brian, 106, 255
knowledge deficit model, 215
Kobrin, Stephen J., 167
Kresge, Naomi, 82
Kuznicki, Jason, 172

Landes, David, 244–45
Laufer, Michael, 84–85
law and lawmakers. *See also* Congress; regulation; soft law
  compromise by special interests, 150–51
  hard vs. soft law, 206, 213, 224–35
  misalignment with social values and civil disobedience, 116–17
  obsolescence and lack of common sense in, 129–31

law of disruption, 153–56. *See also* pacing problem
legal avoidance or indifference vs. civil disobedience, 114
legal defense strategies to support entrepreneurs, 257–29
Lewis, H. W., 40
licensing requirements as brake on innovation, 147–48, 162, 247, 248–50, 256
life expectancy, improvement in, 35
Lindsey, Brink, 143
Liu, Jodie C., 171
lobbying. *See* special interests as barriers to innovation
The Logic of Collective Action (Olson), 145
Longworth, Richard, 30–31
low-information rationality model, 216

maker spaces, 83–84
Maloney, Sean Patrick, 90
Marchant, Gary, 154, 207, 223–24, 239
Maslow's hierarchy of needs, 42–43
Mayhew, Jonathan, 110–11
Mazlish, Bryan, 81
McAfee, Andrew, 191
McCloskey, Deirdre N., 4, 244
McLaughlin, Michael, 49
McLaughlin, Patrick, 133

medical technologies
  biohacking, 79–87
  drones as, 91–92
  mobile, 75–77
  pacing problem with, 127, 155f
  3D-printed prostheses, 78
medical tourism as innovation
    arbitrage, 71–72
medications, DIY, 84–85
MFN. *See* most-favored-nation
    (MFN) clause for technology
    policy
micromobility revolution, 93
Mill, John Stuart, 28
Miller, Henry, 137
Mitchell, Matthew, 145
mobile medical applications,
    75–77
Mokyr, Joel, 243
moonshots vs. iterative
    improvements, 44–47
Moore's Law, 154, 155f
moral dimension of innovation.
    See also civil disobedience
  balancing adaptation with
    values evolution, 185–86
  and economic growth
    imperative, 41–44
  human's innate moral
    sensibility, 194, 195
  importance of freedom to
    humanity, 266–67
  perceived duty to obey the
    law, 110

robotics, 221
  and rule departure by
    government officials,
    124–25
  and soft-law approach, 218–24
  technology as driver of change
    in moral sensibilities,
    194–95
  voluntary ethical codes for
    pacing problem, 223
Morozov, Evgeny, 21
most-favored-nation (MFN)
    clause for technology policy,
    255–56
multistakeholder processes,
    209–11, 222, 227–28
Munemo, Jonathan, 52, 246–47
Murray, Charles, 113, 257
Musk, Elon, 94

Nakamoto, Satoshi, 99
National Highway Traffic Safety
    Administration (NHTSA),
    95–96, 97, 141, 198
nation-state, technology as eroder
    of, 166–68
The Nature of Technology
    (Arthur), 34
Negroponte, Nicholas, 165
NHTSA. *See* National Highway
    Traffic Safety Administra-
    tion (NHTSA)
Nightscout Project, 80–81
Nissenbaum, Helen, 112–13

Noell, Edd S., 41
nongovernmental organizations, technology education role, 218–22
North Carolina State Board of Dental Examiners v. Federal Trade Commission, 248–49
nostalgia, and fear of innovation, 28–29

Obfuscation: A User's Guide for Privacy and Protest (Brunton and Nissenbaum), 112–13
occupational licensing rules, state and local, 147–48, 162, 248–50, 256
offshoring innovation. *See* innovation arbitrage
Olson, Mancur, 145, 245
online predators, fear of, 25
online safety, collaborative regulatory approaches to, 210
OpenAI, 221
OpenBazaar, 100
open-source software, 97–98, 260, 261–62
Ostrom, Elinor, 272
O'Sullivan, Andrea, 179
outcome-based regulation, 208
overcriminalization and regulatory accumulation, 134–35
Owen, Taylor, 166, 204

pacing problem
    as advantage for technology over regulation, 179–80
    Amazon's drone technology and FAA, 68
    contribution to technical civil disobedience, 153–57
    defined, 17
    expected acceleration of, 262
    in medical industry, 127, 155f
    overview, 8–9
    voluntary ethical codes and best practices to solve, 223
pancreas substitute insulin delivery systems, 81–82
Parisi, Francesco, 116–17
Parity Principle for regulation, 256
Partnership on AI, 220–21
people factor in creativity and innovation, 261, 265–66
permissionless innovation
    for born free technologies, 105
    culture of, 243–46
    defined, 17, 49
    development of, 57–58
    and evasive entrepreneurialism, 58–59
    normative case for, 50–52
    protecting the innovator's presumption, 250–52
    and responsible innovation, 196–204
permits and licensing, 147–48, 162, 247, 248–50, 256

Pinker, Steven, 195
Piper, Kelsey, 232
policy. *See* public policy
political entrepreneurialism, 55,
    143–50
The Politics of Innovation
    (Taylor), 143
Pollman, Elizabeth, 59, 118
polycentricity in governance of
    complex systems, 272
Popper, Karl, 269
Posner, Richard, 192, 235
Post, David, 168
Postman, Neil, 188
Potts, Jason, 66
poverty, reduction in, 33f, 35–36
pragmatic optimism, case for,
    267–75
precautionary principle
    defined, 17
    existential threat focus for use
        of, 229–31
    losses of life-saving technology
        caused by, 81–82
    modification of to allow more
        freedom, 198–99
    vs. permissionless approach to
        innovation, 49–51
    rational optimist's mission to
        challenge, 274
    and responsible innovation,
        196–202
prediction markets, distributed,
    101

privacy, right to
    civil disobedience to promote,
        112–13
    digital privacy, 26–27, 170–71
    as source of tech critics'
        concerns, 226–27, 336n20
    surveillance risk from
        precautionary innovation,
        164, 165–66, 170, 232
professional associations,
    technology education role,
    218–22
profit seeking, civil disobedience
    and, 119–22
progress and prosperity of
    civilization. *See* human
    betterment and flourishing
"pro-sumers," 80
psychological vs. physical harms
    from technology, 225
public policy, 241–63. *See also*
    precautionary principle;
    regulation
    and blockchain technology, 102
    call for sensible adjustments to
        innovation, 10–11
    collaborative approach to
        internet policy, 208–10
    communications and media
        policy, 72–73, 136, 146, 251
    defense fund for entrepreneurs
        in relating to, 257–59
    entrepreneurial administra-
        tion, 127–28, 262–63

fear entrepreneurs and false
equivalence in tech debates,
23–28
increasing pressure from
innovators, 10
innovation as priority for,
48–52
for medical devices, 76–77, 81
multistakeholder processes,
209–11, 222, 227–28
pressures of innovation
arbitrage, 73–75
protecting the innovator's
presumption, 250–52
reciprocity agreements to
support innovation, 255–56
recommendations for technology policy, 208
removing barriers to entry for
new ideas, 246–50
respect for consumer choice,
260
and right to earn a living,
250, 256
sunsetting imperative for
regulation, 252–54
purchasing power, increase in,
36–39, 38t
pursuit of happiness, right to, 5–6

quangos (quasi-autonomous
nongovernmental
organizations), 223–24
quantified self, 75, 77

Ranchordás, Sofia, 253–54
rational optimism, case for, 267–75
Rauch, Jonathan, 151
Rawls, John, 114
reciprocity agreements to support
innovation, 255–56
Rees-Mogg, William, 67
regulation. *See also* pacing
problem; technological
governance
accumulation of and rise of
resistance, 132–35
alternative transportation
challenges for, 59–64, 70,
92–94, 178–79
appropriate role of, 123
beneficiaries' reluctance to give
up power from, 247
biohacking's challenge to, 87
compromising with
regulators, 63
continued proliferation of, 160
cryptocurrencies, 99–100, 172
evasive food entrepreneurs
and, 88–89
failure of reforms to constrain,
160–62
as fear response, 20
effect of innovation arbitrage
on, 73
innovation as constant
pressure on, 173
lack of common sense in,
129–31

lack of update and adjustment, 133–34, 134f, 256

on micromobility revolution, 93

need for innovative and flexible approach, 204, 263

Parity Principle for, 256

pros and cons of exerting control over innovation, 185

regulators as risk educators, 216

spontaneous deregulation, 103–4

sunsetting imperative for, 252–54

regulatory capture, 51, 61–62, 65, 69, 148–49, 161

regulatory entrepreneurs, 17, 54–57, 59–65

regulatory sandboxes, 208

Reimschisel, Jordan, 77

rent-seeking behavior, 55, 143, 146

responsible innovation, 196–204, 211–12, 213, 214–15

responsible research and innovation (RRI), 196–204, 213, 214–15

RFID (radio frequency identification) chips, implantable, 85–86

ridesharing services, 59–64, 70, 93, 178

Ridley, Matt, 27–28, 267

rights debate between tech promoters and tech critics, 226–27

right to earn living and innovate, 5–6, 249–50, 256

risk education, 215–18

risks of new technology

existential risks, 228–36

hard-law vs. soft-law approaches, 206, 213, 224–28

prioritizing, 235–40

reality of, 20

risk taking

in civil disobedience, 121–22

dampening effect of risk-averse regulation, 68

as drivers of progress and human flourishing, 3–13, 51

as necessary to innovation, 40–44

risk-weighted regulation, 208

robotics, ethical codes and international agreements, 221

rogue satellite launches, as innovation arbitrage, 72–73

Romer, Paul, 181

Roth, Alvin E., 189

Rowes, Jeff, 143

RRI. *See* responsible research and innovation (RRI)

rule departure by government, 103, 124–28

The Rule of Nobody (Howard), 132

Ruling the Waves: Cycles of Discovery, Chaos, and Wealth from the Compass to the Internet (Spar), 179

Sacasas, L. M., 190–91

SatoshiDice, 305n252

Schomberg, René, 196, 203

Schumpeter, Joseph, 39, 56, 143

Scott, James C., 271–72

Scribner, Marc, 141

Securities and Exchange Commission (SEC), 101

semi-autonomous vehicles, 94–97

Shirky, Clay, 168–69

Silver, Adam, 101

Simon, Julian, 261

Skorup, Brent, 251

smart glasses, initial failure of, 189–90

smartphones, multiple features of, 39

SmileDirectClub, 83

Smith, Adam, 193–94

Smith, Stephen L. S., 41

social entrepreneurialism, 47–48. *See also* free innovation

social justice motivation for civil disobedience, 112–13

socially responsible innovation, 196–204, 211–12, 213, 214–15

social networking safety, 25

social norms, using to maintain industry accountability for technology development, 222–23

social welfare. *See* human betterment and flourishing

societal effect of innovation, 5, 265–66. *See also* economic growth; moral dimension of innovation

soft determinism on technology, 190–91

soft law, 205–40
basic tenets of, 206–15
challenge of defining harm, 224–28
defined, 17
and driverless cars, 95–96
existential risks and need for hard law, 228–35
in FDA's approach to digital health, 76–77
hard-law approach vs., 11–12
importance for adapting regulation, 263
professional associations and ethical codes, 218–24
and responsible innovation, 199
risk education, 215–18
risk prioritization, 235–40

Spar, Debora L., 179

special interests, as barriers to innovation, 149, 160, 241–42, 248–49

spontaneous deregulation, 103–4

sports gambling, 101

Starship Technologies, 71

state secrets, leaking of, 169–70

state vs. federal power, 16, 70–71, 161

status quo attachment, 20, 136

Stone, Brad, 61

Sunsetting Imperative, 252–54

surveillance risk from precautionary innovation, 164, 165–66, 170, 232

Swarm Technologies, 72–73

synthetic biology, 233

Tarr, Nina W., 88–89

tax policies, and cryptocurrencies, 100

Taylor, Mark Zachary, 143

tech critics (techno-pessimists), critique of, 186–93, 212, 236, 242, 268–71, 274

technological civil disobedience
  in Bitcoin startup, 99
  characteristics of, 114
  defined, 2, 17
  ex ante vs. ex post question for entrepreneurs, 122–24
  freedom of association and spirit of resistance, 115–17
  libertarian motivation for, 113
  and profit seeking, 119–22
  and rule departure, 124–28

technological governance. *See also* permissionless innovation; regulation; soft law
  blockchain as form of, 102
  common-sense approach to, 10–11

  and social vs. commercial entrepreneurialism, 47–48
  virtue of bottom-up approach, 271–72, 272t, 274

technological harm, theory of, 224–28

technologies of freedom, 2, 17, 154–56, 168–73. *See also* permissionless innovation

technology. *See also* innovation
  born free or born captive to regulation, 104–7, 138–39
  as challenger to government power, 168–75
  as dependent on human motivations and activity, 189, 192–93
  disruption by internet and other digital tech, 166–67
  empowerment of government by, 163–65
  as eroder of nation-state, 166–68
  opposition to new, 21–28
  pace of change in, 8–9
  role in life improvements, 45f

Technology and the End of Authority (Kuznicki), 172

Teles, Steven, 143, 162

Tesla electric vehicles, 94–95, 144

Theory of Moral Sentiments (Adam Smith), 193–94

Theranos, 123

Thiel, Peter, 39

Thompson, Clive, 27–28

Thoreau, Henry David, 111

threat inflation, 24

3D printing, 77–79, 128, 156

Tibbetts, Jeffrey, 85–86

Toffler, Alvin, 23

too-big-to-ban regulatory
    entrepreneurialism
    strategy, 61

Topol, Eric, 75

traffic-monitoring applications,
    92–93

transparency laws in technological
    governance, 239

transportation
    alternative types that challenge
        regulation, 59–64, 70, 92–94,
        178–79
    driverless cars, 70, 94–98, 127,
        128, 139, 309n12
    regulatory capture and
        evasive entrepreneurialism,
        61–62, 139

Travis's Law, 61

Tuccille, J. D., 78

Tusk, Bradley, 60, 64–65, 259

23andMe, 65, 69

Tyler, Tom R., 110

UAS (unmanned aircraft system).
    See drones

Uber (ridesharing service),
    59–64, 70, 178

Unleash Your Inner Company
    (Chisholm), 250

unmanned aircraft system (UAS).
    See drones

upstream governance, 196

user entrepreneurialism, 47, 48

utopianism vs. pragmatic change,
    12–13, 34, 163, 267–69

vaccines, development of, 36

Valdivia, Walter D., 200

voice and exit, innovation as, 13,
    114–15, 175–79

voluntary association, civil
    disobedience as exercise in,
    115–16

von Hippel, Eric, 47

Wallach, Wendell, 153, 207,
    223–24, 239

Wallison, Peter J., 151

Walton, Douglas, 237

Wangenheim, Georg von,
    116–17

Waze, 92–93

*The Wealth and Poverty of
    Nations: Why Some Are So
    Rich and Some Are So Poor*
    (Landes), 244–45

Weapons of Mass Destruction
    (WMDs), 236–37

Webb, Bruce G., 41

Weiser, Philip, 127, 204, 207

wellness tech, 75

Why People Obey the Law
(Tyler), 110
Williams, Richard, 263
Wilson, James Q., 136, 313n74
Wilt, Michael, 133
Wittes, Benjamin, 171

WMDs. *See* Weapons of Mass
Destruction (WMDs)
Wooldridge, Adrian, 244
Wyatt, Sally, 187–88

Zinn, Howard, 112

Adam Thierer is a senior research fellow at the Mercatus Center at George Mason University. He has spent more than 25 years covering the intersection of emerging technologies and public policy and has authored or edited eight books on topics ranging from media regulation and child safety issues to the role of federalism in high-technology markets. His previous book was *Permissionless Innovation: The Continuing Case for Comprehensive Technological Freedom*.

His writings have appeared in the *Wall Street Journal*, *The Economist*, the *Washington Post*, and *The Atlantic*, and he has testified numerous times on Capitol Hill, before regulatory agencies, and in state legislatures. Thierer has also served on several distinguished online safety task forces, including Harvard University's Internet Safety Technical Task Force and the Obama administration's Online Safety Technology Working Group.

Previously, Thierer was president of the Progress and Freedom Foundation, director of Telecommunications Studies at the Cato Institute, and a senior fellow at the Heritage Foundation. He received his MA in international business management and trade theory at the University of Maryland and his BA in journalism and political science from Indiana University.

Founded in 1977, the Cato Institute is a public policy research foundation dedicated to broadening the parameters of policy debate to allow consideration of more options that are consistent with the principles of limited government, individual liberty, and peace. To that end, the Institute strives to achieve greater involvement of the intelligent, concerned lay public in questions of policy and the proper role of government.

The Institute is named for *Cato's Letters*, libertarian pamphlets that were widely read in the American Colonies in the early 18th century and played a major role in laying the philosophical foundation for the American Revolution.

Despite the achievement of the nation's Founders, today virtually no aspect of life is free from government encroachment. A pervasive intolerance for individual rights is shown by government's arbitrary intrusions into private economic transactions and its disregard for civil liberties. And while freedom around the globe has notably increased in the past several decades, many countries have moved in the opposite direction, and most governments still do not respect or safeguard the wide range of civil and economic liberties.

To address those issues, the Cato Institute undertakes an extensive publications program on the complete spectrum of policy issues. Books, monographs, and shorter studies are commissioned to examine the federal budget, Social Security, regulation, military spending, international trade, and myriad other issues. Major policy conferences are held throughout the year, from which papers are published thrice yearly in the *Cato Journal*. The Institute also publishes the quarterly magazine *Regulation*.

In order to maintain its independence, the Cato Institute accepts no government funding. Contributions are received from foundations, corporations, and individuals, and other revenue is generated from the sale of publications. The Institute is a non-profit, tax-exempt, educational foundation under Section 501(c)3 of the Internal Revenue Code.

CATO INSTITUTE
1000 Massachusetts Avenue NW
Washington, DC 20001
www.cato.org